South University Library
Richmond Campus
2151 Old Brick Road
Glen Allen, Va 23060

Digital Solidarities, Communication Policy and Multi-stakeholder Global Governance

This book is part of the Peter Lang Media and Communication list.
Every volume is peer reviewed and meets
the highest quality standards for content and production.

PETER LANG
New York • Washington, D.C./Baltimore • Bern
Frankfurt • Berlin • Brussels • Vienna • Oxford

Marc Raboy, Normand Landry, Jeremy Shtern

Digital Solidarities, Communication Policy and Multi-stakeholder Global Governance

The Legacy of the World Summit on the Information Society

PETER LANG
New York • Washington, D.C./Baltimore • Bern
Frankfurt • Berlin • Brussels • Vienna • Oxford

Library of Congress Cataloging-in-Publication Data
Raboy, Marc.
Digital solidarities, communication policy and multi-stakeholder
global governance: the legacy of the World Summit on the Information Society /
Marc Raboy, Normand Landry, Jeremy Shtern.
p. cm.
Includes bibliographical references and index.
1. World Summit on the Information Society. 2. Information society.
3. Communication policy—International cooperation.
4. Telecommunication policy—International cooperation.
5. Information technology—International cooperation.
6. Civil society. I. Landry, Normand. II. Shtern, Jeremy. III. Title.
HM851.R333   303.48'33090511—dc22   2010037544
ISBN 978-1-4331-0740-5

Bibliographic information published by **Die Deutsche Nationalbibliothek**.
Die Deutsche Nationalbibliothek lists this publication in the "Deutsche
Nationalbibliografie"; detailed bibliographic data is available
on the Internet at http://dnb.d-nb.de/.

The paper in this book meets the guidelines for permanence and durability
of the Committee on Production Guidelines for Book Longevity
of the Council of Library Resources.

© 2010 Peter Lang Publishing, Inc., New York
29 Broadway, 18th floor, New York, NY 10006
www.peterlang.com

All rights reserved.
Reprint or reproduction, even partially, in all forms such as microfilm,
xerography, microfiche, microcard, and offset strictly prohibited.

Printed in Germany

Contents

List of Figures ix
List of Tables xi
List of Acronyms xiii
Acknowledgments xvii
Introduction 1
    The UN World Summit on the Information Society (WSIS) 2

Part One: A Summit in Context

Chapter One. A Summit in Two Phases/A Two-phased Summit 11
    A Summit in Two Phases 11
    What Was New, Different, and the Same at Phase II 14
    Overview of Themes and Issues Discussed at WSIS 17
    Civil Society at WSIS: Backgrounds, Structures and Practices 31
    Civil Society Structures at the WSIS 34

Chapter Two. Advancing Through the Phase II Preparatory Framework 41
    December 2003–June 2004: Reconstruction and Remobilization 43
    PrepCom I: Hammamet, 24-26 June 2004 47
    A First Interlude: June 27, 2004-February 16, 2005 53
    PrepCom II: 17-25 February 2005, Geneva (Switzerland) 59
    A Second Interlude: February 26, 2005-September 18, 2005 64
    PrepCom III: September 19-30, 2005, Geneva (Switzerland),
        and November 13-15, 2005, Tunis (Tunisia) 68
    PrepCom III Resumed:
        The Final Countdown, November 13-15, 2005 (Tunis) 73
    Is This the End? The Tunis Summit, 16-18 November, 2005 76

Chapter Three. Civil Society at WSIS Phase II: A Summary Assessment 79
    New Phase, New Context, New Structures? 79

CS Refuses to Remove Its Foot from the Doorway:
   The Campaign for Multi-stakeholder Global Governance at Phase II 88

Part Two: WSIS Phase II Issues and Outcomes

Chapter Four. Digital Solidarity?
   Financing Access to the Information Society 101
      Financing at WSIS II: Issues and Controversies 102
      The Question of Financing at WSIS: The Path Chosen by the
         International Community 112
      Assessing Financial Mechanisms: Civil Society's Positions 115

Chapter Five. A Geopolitics of Networks: Internet Governance at WSIS 119
      Internet Governance at WSIS Phase I:
         [International/Intergovernmental] 120
      The Working Group on Internet Governance (WGIG) 131
      From ICANN to Internet Governance 135
      Internet Governance at WSIS Round III: The Tunis Compromise 140

Chapter Six. Implementation and Follow-up 149
      The WSIS as a Test for Implementation and Follow-up? 149
      Civil Society's Responsibilities and Challenges 152
      WSIS Implementation and Follow-up: An Overview 159
      Implementing Lessons Learned 165

Part Three: Civil Society and Global Communication Governance Beyond WSIS

Chapter Seven. Civil Society, Internet Governance and the IGF 169
      Addressing the Democratic Deficit in CS? The Internet Governance
         Caucus Charter Drafting/Adoption Process. 172
      Internet Governance Forum (IGF) 179
      The IGF and Multi-stakeholder Global Governance 187
      The Multi-stakeholder Model:
         Reflections from the Internet Governance Experience 191

Chapter Eight. Post-WSIS Civil Society Engagement 201
    WSIS Implementation and Follow-up 202
    UNESCO 204
    World Social Forum 205
    ICANN 206
    OECD 208
    WIPO 209
    GAID 211
    ITU 212

Chapter Nine. Multi-stakeholder Global Governance
  at the WSIS and Beyond 217
    The Substantive Legacy of the WSIS 217
    CS at WSIS in Context and Conceptions
      of Multi-stakeholder Democracy 229
    Models of Global Governance 232

Appendix: "Much More Could Have Been Achieved" 237
Bibliography 259
Index 269

Figures

Figure 1: The WSIS Process.. 12
Figure 2: Outline of the Preparatory Process for the Tunis Phase of WSIS 48
Figure 3: Post-Tunis Mechanisms Allowing for NGO Inputs 160

Tables

Table 1: Detailed Timetable of Official Events and Preparatory Process 42
Table 2: Definitions of and Guidelines for CS Structures 86
Table 3: Summary- The WGIG Models for Institutional Reform of Global Internet Governance 137
Table 4: IGF Meeting Themes: 2006-2009 183

Acronyms

AMARC	World Association of Community Radio Broadcasters
APC	Association for Progressive Communications
C & T	Content and Themes
CCBI	Coordinating Committee of Business Interlocutors
CCTLD	Country Code Top Level Domain Name
CEB	Chief Executives Board for coordination
CONGO	Conference of Non-Governmental Organizations in Consultative Relationship with the United Nations
CRIS	Communication Rights in the Information Society
CS	Civil Society
CSB	Civil Society Bureau
CSD	Civil Society Division
CSO(s)	Civil Society Organisation(s)
CSP	Civil Society Plenary
CSTD	Commission on Science and Technology for Development
CTG:	Content and Themes Group
DC(s)	Dynamic Coalition(s)
DNS	Domain Name System
DSF	Digital Solidarity Fund
ECLAC	Economic Commission for Latin America and the Caribbean
ECOSOC	United Nations Economic and Social Council
EU	European Union
FAO	Food and Agriculture Organization
G8	Group of Eight
G20	Group of Twenty
GAC	Governmental Advisory Committee (ICANN)
GAID	Global Alliance for ICT and Development

GFC	Group of Friends of the Chair
GNGO(s)	Governmental Non-Governmental Organization(s)
GTLD	Generic Top-Level Domain
HLSOC	High Level Summit Organizing Committee
HRC	Human Rights in China
IAB	Internet Architecture Board
IAEA	International Atomic Energy Agency
IANA	Inernet Assigned Numbers Authority
ICANN	Internet Corporation for Assigned Names and Numbers
ICD	Information and Communications for Development
ICTs	Information and Communication Technologies
ICTD	Information and Communication Technology for Development
IETF	Internet Engineering Task Force
IG	Internet Governance
IGC	Internet Governance Caucus
IGF	Internet Governance Forum
ILO	International Labour Organization
IP	Internet Protocol
IPR	Intellectual Property Rights
ISOC	Internet Society
ISP(s)	Internet Service Provider(s)
ITU	International Telecommunication Union
JPA	Joint Project Agreement
MAG	Multi-stakeholder Advisory Group
MDGs	Millennium Development Goals
MoU	Memorandum of Understanding
NCUC	Non Commercial Users Constituency (ICANN)
NGO(s)	Non-Governmental Organization(s)
NICTs	Networked Information and Communication Technologies

NTIA	National Telecommunications and Information Administration
ODA	Official Development Assistance
OECD	Organization for Economic Co-operation and Development
PPP	Public Private Partnership
PrepCom	Preparatory Committee
RIR	Regional Internet Registry
TFFM	Task Force on Financial Mechanisms
TRIPS	Trade-Related Aspects of Intellectual Property Rights
UDHR	Universal Declaration of Human Rights
UK	United Kingdom
UN	United Nations
UNCTAD	United Nations Conference on Trade and Development
UNDESA	United Nations Department of Economic and Social Affairs
UNDP	United Nations Development Programme
UNECA	United Nations Economic Commission for Africa
UNECE	United Nations Economic Commission for Europe
UNESCO	United Nations Educational, Scientific and Cultural Organization
UNESCWA	United Nations Economic and Social Commission for Western Asia
UNGA	United Nations General Assembly
UNGIS	United Nations Group on the Information Society
UNHCHR	United Nations High Commissioner for Human Rights
UNHCR	United Nations High Commissioner for Refugees
UNICEF	United Nations Children's Fund
UNIDO	United Nations Industrial Development Organization
UNITAR	United Nations Institute for Training and Research
UNODC	United Nations Office on Drugs and Crime
UNRWA	United Nations Relief and Works Agency

UNSG	United Nations Secretary General
UNWTO	United Nations World Tourism Organization
UPU	Universal Postal Union
US	United States
US DOC	United States Department of Commerce
WB	World Bank
WFP	World Food Programme
WG	Working Group
WGIG	Working Group on Internet Governance
WHO	World Health Organization
WIPO	World Intellectual Property Organization
WMWG	Working Methods Working Group
WMO	World Meteorological Organization
WSIS	World Summit on the Information Society
WSIS ES	World Summit on the Information Society Executive Secretariat
WTO	World Trade Organization

Acknowledgments

We wish to thank Juliana Santos Botelho for producing an exhaustive synthesis of civil society listserv exchanges from Phase II of the WSIS, which got us started.

Oleh Replansky translated a substantial part of the first draft of our manuscript which was originally written in French.

The Beaverbrook Canadian Foundation and Media@McGill provided crucial support towards the costs of translation and indexing.

At Peter Lang, our commissioning editor, Mary Savigar, was an endless source of encouragement and Sophie Appel oversaw the design and production with skill and grace. We were unable to exhaust their patience, however hard we tried.

Above all, we thank the hundreds of WSIS civil society activists whose dedication to the democratization of global communication governance inspired this book.

M.R.
N.L.
J.S.
Montréal, May 2010

Introduction

In 2003 and then again in 2005, the international community was called by the United Nations to take part in a World Summit on the Information Society (WSIS). This two-phased UN Summit placed an unprecedented global spotlight on information and communication issues. It was also a grand experiment in global governance, including (to an arguably ground-breaking extent) the active participation of non-governmental stakeholders in the development of public policies at the international level.

An earlier book by two of the authors of the present volume (Marc Raboy and Normand Landry, *Civil Society, Communication and Global Governance: Issues from the World Summit on the Information Society* (Peter Lang, 2005) provided a sweeping portrait of the players, structures and themes of the WSIS as well as a critical analysis of the summit's first phase. Its particular focus was on the issues raised and the role played by 'civil society.' According to the London School of Economics' Centre for Civil Society

> Civil society refers to the arena of uncoerced collective action around shared interests, purposes and values. In theory, its institutional forms are distinct from those of the state, family and market, though in practice, the boundaries between state, civil society, family and market are often complex, blurred and negotiated. Civil society commonly embraces a diversity of spaces, actors and institutional forms, varying in their degree of formality, autonomy and power. Civil societies are often populated by organisations such as registered charities, development non-governmental organisations, community groups, women's organisations, faith-based organisations, professional associations, trades unions, self-help groups, social movements, business associations, coalitions and advocacy groups.[1]

Though this definition is arguably more developed and nuanced than the context-specific understanding of civil society that emerged over the course of the WSIS, in this book we will primarily use the term civil society to describe the NGOs, activists, academics and other non-business, non-governmental stakeholders who, during the first phase of the WSIS, were granted precedent-setting participatory access to UN policymaking.

There is no doubt that Phase I of the WSIS was a watershed moment in respect to global communication policymaking and the institutional frame-

---

1 LSE Centre for Civil Society, *What Is Civil Society* (April 23, 2009). http://www.lse.ac.uk/collections/CCS/introduction/what_is_civil_society.htm

work of international politics. However, *Civil Society, Communication and Global Governance* covered but the first act of an emerging phenomenon rather than its entire story. As the focus of the WSIS shifted away from the more general task of producing a series of principles to guide the emergence of a global 'information society' to actually negotiating specific agreements on fundamental questions such as financial solutions to the digital divide and Internet governance, a different and more vivid picture of the impact of the WSIS on global communication policy and on the ideals of multi-stakeholder governance took shape. Over the course of WSIS Phase II, different actors would take centre stage and the structures and processes discussed in *Civil Society, Communication and Global Governance* would continue to develop and to be subjected to criticism. As a result, the second phase of the WSIS and the legacy of the institutions and partnerships that it has created is a rich and vibrant case study in the institutional innovations that are shaping not only how global communication is being governed but even the very notion of who governs the information society.

Digital Solidarities, Communication Policy and Multi-stakeholder Global Governance: The Legacy of the World Summit on the Information Society picks up where the previous volume left off. It examines the distinct players, structures and themes of the second phase of the WSIS, once again with a particular focus on the issues raised and roles played by civil society. It includes discussion of the Internet Governance Forum—the new multi-stakeholder organization created as the most tangible output of the WSIS—as well as discussion of how the process of civil society self-organization has continued post-WSIS to reflect on the entirety of the WSIS experience and what it tells us about the challenges and opportunities embedded in the notion of multi-stakeholder governance as well as on what globalization and international politics now mean to the global governance of communication.

The UN World Summit on the Information Society (WSIS)

By adopting resolution A/RES/56/183[2] on December 21, 2001, the General Assembly of the United Nations officially set its feet on uncharted territory in regard to the organization of UN Summits. The Assembly was "convinced of the need, at the highest political level, to marshal the global consensus and

---

2   See United Nations General Assembly, *Resolution A/RES/56/183, World Summit on the Information Society.* (January 31, 2002). http://www.itu.int/wsis/docs/background/resolutions/56_183_unga_2002.pdf

commitment required to promote the urgently needed access of all countries to information, knowledge, and communication technologies for development." In response, the UNGA gave the International Telecommunication Union (ITU) the mandate to take the lead on organization of a World Summit on the Information Society.

There are four major elements that distinguished the WSIS process from typical UN World Summit practice: the nature of the topics under discussion, the participatory processes that were established, the division of the summit into two distinct phases, and the effort to establish implementation and follow-up mechanisms as part of the negotiations held at the summit itself.

The WSIS was the first event of its kind in the history of the United Nations to be devoted exclusively to issues of information and communication.[3] Its convening suggested an interest, at the highest levels of intergovernmental politics, to put social, economic, and cultural development issues intertwined with communication and knowledge sharing onto the agenda of the United Nations. The WSIS also demonstrated a consensus that effective policy responses to such issues need to bridge traditional national and state governmental frameworks. The technological innovations of the last few decades having already weakened (although by no means having destroyed) the ability of states to control the flow of information within or outside of their borders, the sentiment was that it had become necessary to work in a multi-stakeholder international context in order to properly address such issues. One premise of the WSIS was, in other words, that communication governance can only be effective if political, economic, and social actors work together as partners. Its convening suggested rising acceptance that such policy responses necessitate that the governance of communication be gradually displaced from the na-

---

3   Prior to the WSIS, there had been a series of previous high level intergovernmental events that focused on information and communication including the 1948 UN Conference on Freedom of Information that was convened to contribute to the drafting of the Universal Declaration of Human Rights (UDHR) and the "Right to Communicate" discussions held at UNESCO in the 1970s and 1980s which culminated in the creation of an "International Commission for the Study of Communication Problems" (often referred to as the MacBride Commission). Discussions of media, communication and information issues at the intergovernmental level proved, in each case, controversial, politically charged and ultimately fraught for the host organization. Debates at the 1948 Conference on Freedom of Information were highly polarized and, beyond inclusion of "freedom of expression" as eventual Article 19 in the UDHR, most of the proposed outcomes were dropped for lack of anything resembling consensus support. Contempt for the report produced by the MacBride Commission was so strong that the US, UK and Singapore pulled out of UNESCO in its aftermath, striking a crippling blow to the credibility and funding of the organization (see Marc Raboy and Jeremy Shtern, *Media Divides: Communication Rights and the Right to Communicate in Canada*. Vancouver: UBC Press, 2010, Ch. 2).

tional to the highly specialized international level.[4] From the outset, the WSIS was framed as a political initiative undertaken in response to an unprecedented technological revolution. Its goal, according to summit organizers, was to orient the benefits of this revolution toward global social and economic development.[5]

The WSIS was meant to address a set of issues associated with the rise of a globally networked society marked by strong factors of social, cultural, and economic exclusion. Developed in conjunction with the UN Millennium Development Goals,[6] which focus on the eradication of poverty, its official goals also aimed at orienting the benefits of new information and communication technologies towards international development. This implied above all that participating states should agree on a common vision for the information society, encourage communication-based infrastructure development, develop human resources and knowledge, ensure financial and knowledge transfers to developing countries, and enhance linguistic and cultural diversity using new communication and information technologies.

Regardless of these official *raisons d'être*, however, several noticeably different themes permeated and monopolized a large portion of the negotiations that took place in Geneva and Tunis. Issues surrounding human rights, the global intellectual property regime, the role of the so-called "traditional" media in the information society, Internet governance, and the marginalization of vulnerable groups complicated the negotiation process. As a result, the first phase of the WSIS was dedicated to defining the scope of the themes being discussed within the framework of the summit. Non-governmental organizations and other civil society actors present at the event worked to redirect political discussions around themes of inclusion and social partnership, as well as linguistic, cultural, and sexual diversity, that would create a base for a free and inclusive information society out of the international normative framework for human rights. These civil society organizations eventually presented a common vision by jointly drafting a declaration of principles that outlined the values upon which an information and communication society should be based.

---

4 This can be seen most notably in the fields of intellectual property rights (WIPO), commercial agreements (WTO), culture and cultural diversity (UNESCO), and technological standardization and radio airwave ownership management (ITU). Each of these institutions has its own regulations, practices, dynamics, and mandates.

5 See the ITU Press Office, *Framework and Venue of the World Summit on the Information Society Announced.* (June 8, 2001). http://www.itu.int/newsarchive/press_releases/2001/12.html

6 See http://www.un.org/millenniumgoals/

The declaration, entitled *Shaping Information Societies for Human Needs*, was accepted as an official document at the Geneva Summit in December 2003.[7]

The multi-stakeholder aspect of the WSIS also made the conference a sort of testing ground for international governance processes, as the summit became a site for redefining the roles and responsibilities of non-governmental actors in the development of supranational politics. The multi-stakeholder aspect of the WSIS will therefore be discussed, analyzed, and critiqued in detail throughout this work.

The category of "NGO in consultative status with the UN" has existed in the UN system for decades and is a general framework used to accredit organizations and set guidelines for stakeholder participation within the UN system. However, binding, uniform rules outlining the processes through which specific agencies, programs and events accredit organizations and define the modalities of their participation did not exist at the time of the convening of the WSIS.[8] Consequently, UN events such as World Summits are required to establish their own internal regulations for the participation of stakeholder groups. This means that political negotiations must be held during the preparatory stages of such events in order to determine and agree upon the terms of participation to be applied to all parties. The result of such negotiation processes is that the regulations established reflect not only political interests, but also modifications, innovations and efforts to reform conventions in general UN governance practices. As they take shape, in other words, the regulations reveal new trends in the way that international governance is conducted. During its first gathering in July 2002, the WSIS preparatory committee (PrepCom) agreed on a set of internal rules that cleared the way for the participation of all governmental and non-governmental political actors.[9]

---

7 The document is available online at http://www.itu.int/wsis/docs/geneva/civil-society-declaration.pdf
8 In response to perceptions that the importance of NGOs within the UN system was increasing, the UN Secretary-General appointed a panel of eminent persons to study the issue. Chaired by the former president of Brazil, Fernando Henrique Cardoso, the panel spent a year taking stock of existing practice, consulting widely with interested parties and proposing better ways of managing United Nations–civil society relations, releasing its report in June 2004. The need to establish such uniform rules and proposals for the accreditation of NGOs to the UN were topics of considerable discussion within the report of the Cardoso Committee. See Fernando Henrique Cardoso et al., *We the Peoples: Civil Society, the United Nations and Global Governance. Report of the Panel of Eminent Persons on United Nations–Civil Society Relations* (A/58/817). (June 11, 2004). http://www.un.org/french/ga/search/view_doc.asp?symbol=A/58/817&referer=http://www.un.org/french/reform/panel.html&Lang=E
9 The internal regulation adopted during the first PrepCom of the Geneva phase was also used for the second phase. See the WSIS Executive Secretariat, *Report of the First Meeting of the Prepara-*

Civil society participation in the WSIS was therefore regulated by various procedures and practices. Some of these regulations were followed vigorously and consistently, while others were often omitted or even completely ignored. Political contingencies played an essential role in determining the level of inclusion for the non-governmental organizations participating in the summit.

The WSIS was all the more atypical for UN process in that the General Assembly called for a multi-stakeholder summit to be held in two phases and in two distinct locations. The first phase of the WSIS, which ended in Geneva in December 2003, was thoroughly described and analysed in *Civil Society, Communication, and Global Governance*. The second phase of the WSIS, also known as the Tunis phase, consisted of preparatory stages that took place in both Switzerland and Tunisia and culminated with a summit event held in Tunis in November 2005. In this book we turn our attention to the second phase of the WSIS and its aftermath and reflect on the legacy of the entire WSIS experience for the global governance of communication.

The WSIS opened in Switzerland, a developed country of the North, globally renowned for hosting high-level international political conferences. It later moved to Tunisia, a North African country with no such tradition. This in itself was an innovative approach; nevertheless, the discussions held in Tunis were more focused than those held in Geneva. The Tunis phase was primarily devoted to consolidating gains and solving disputes that had carried over from the first phase and elucidating first-phase decisions into concrete initiatives. Furthermore, the selection of Tunisia as a host country for the WSIS was strongly contested by various actors due to the country's uneven performance in regard to respecting and protecting human rights and fundamental freedoms.

In parallel to the first phase of the WSIS, the UN formed an Ad Hoc Working Group of the General Assembly to examine the "integrated and coordinated implementation of and follow-up to the outcomes of the major United Nations conferences and summits in the economic and social fields." The report of this Working Group determined that "progress in implementation has been insufficient and therefore the time has come to vigorously pursue effective implementation." In response, UN General Assembly resolution 57/270 was passed in July of 2003, emphasizing that

> the United Nations system has an important responsibility to assist Governments to stay fully engaged in the follow-up to and implementation of agreements and commitments reached at the major United Nations conferences and Summits, and invites

---

tory Committee (WSIS03/PREP-1/11(Rev.1)-E). (July 12, 2002). http://www.itu.int/dms_pub/itu-s/md/02/wsispc1/doc/S02-WSISPC1-DOC-0011!R1!MSW-E.doc

its intergovernmental bodies to further promote the implementation of the outcomes of the major United Nations conferences and Summits (57/270 at para 6).

With the first phase of the WSIS entering its final stages as this resolution was being passed, the second phase obviously emerged as an opportunity for the UN to demonstrate this newly minted commitment by building plans for its own implementation and follow-up into the outcomes of the second phase.

The distinct nature of the WSIS was thus established by the confluence of a series of general political trends and desires for innovation in UN governance practice that combined to position this particular summit as a unique political experience that would include new actors, test new participation and consultation mechanisms, and leverage the possible advantages of a two-phased political negotiation process.

PART ONE
A Summit in Context

The organization of the WSIS took place in a context marked by multiple political, organizational and thematic contingencies. As a multi-stakeholder experiment organized in two distinct phases, the WSIS was to serve both as a framework for negotiations on very specific political issues and as a test for a more inclusive global governance model. This dual mandate presented itself as a considerable challenge for the organizers of the event.

In this first part we provide the foundations of this book. Part One will present the political and institutional context in which both the conclusion of the first phase and the organization of the second phase of the WSIS took place. It will further detail the institutional and organizational structures of the WSIS as well as the themes and issues addressed at the event.

Within this larger narrative and overview, we will focus in particular on the participation of civil society at the WSIS. Part One will detail the various positions developed by civil society organizations on the major themes and issues of the summit, address problems of organization, participation and inclusion faced by civil society during WSIS Phase II, and present CS assessments of the politics and issues shaping the intergovernmental negotiations.

• CHAPTER ONE •

A Summit in Two Phases/ A Two-phased Summit

A Summit in Two Phases

In this chapter, we will look at the different dimensions of the summit's organizational structure. The first phase of the WSIS ended in Geneva on December 12, 2003, with the adoption of a Declaration of Principles and a Plan of Action. Three planned meetings of the preparatory committee, a series of hastily organized additional PrepComs, five regional conferences, and a series of themed preliminary meetings and sessions were required for this consensus to be reached in Geneva.[1]

The acceptance of an agreement—reached *in extremis*—encompassing nearly all of the issues debated during the first phase of the WSIS, averted the UN system from a political failure that had seemed inevitable to many observers and participants in the lead up to the summit. However, the Geneva phase exposed deep divisions between states in regard to certain key issues. Three issues, in particular, were not solved during the first phase and were therefore deferred to the second: Internet governance, financial mechanisms for eradicating the so-called "digital divide," and implementation and follow-up of summit outcomes. These three issues set the substantive agenda of the WSIS's Tunis phase.

---

1   The story of Phase I of the WSIS is the object of the companion volume to the present book. See Marc Raboy and Normand Landry, *Civil Society, Communication and Global Governance: Issues from the World Summit on the Information Society*. New York: Peter Lang, 2005.

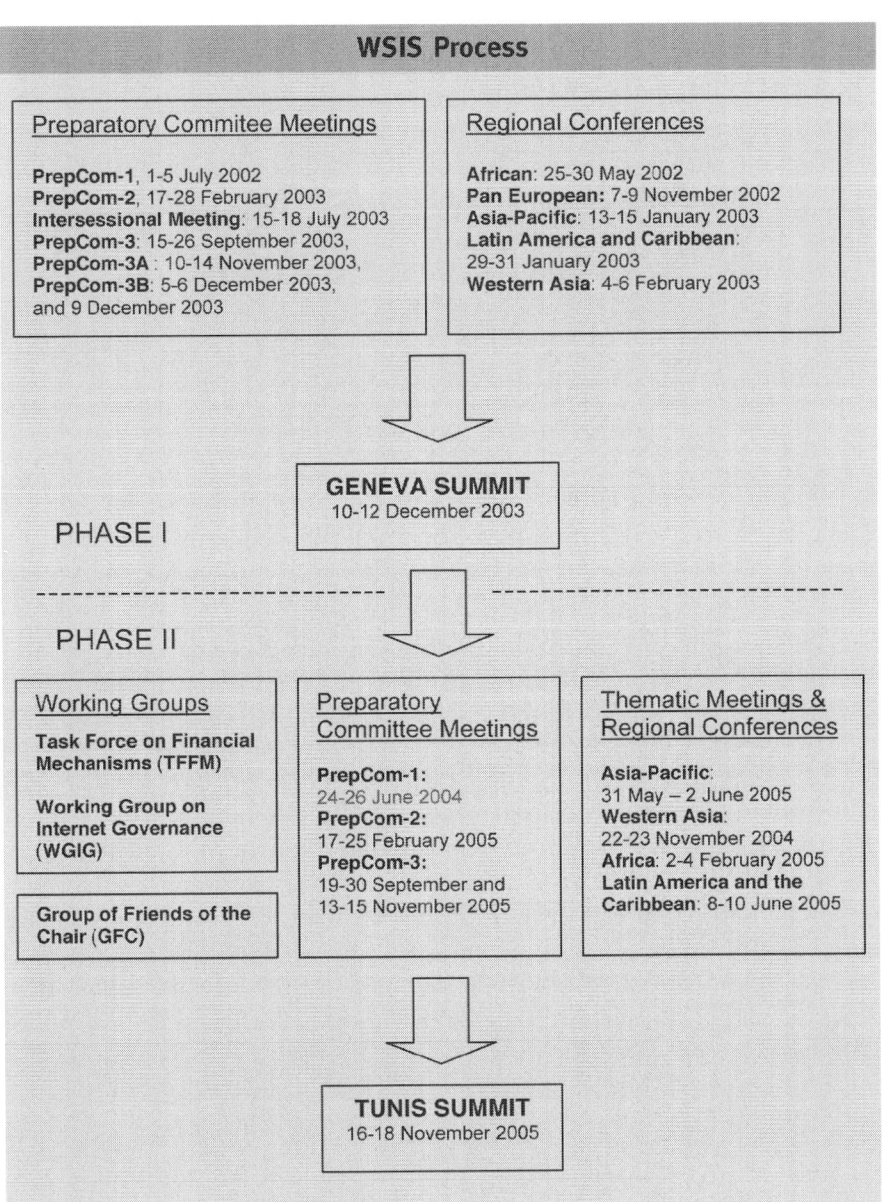

Figure 1: The WSIS Process[2]

---

2   Conference of Non-Governmental Organizations in Consultative Relationship with the United Nations (CONGO), *Civil Society Orientation Kit*. (November 2005). p. 5. http://www.ngocongo.org/congo/files/wsis_oriention_kit.pdf

The preparatory process for the second phase took place in Tunisia and Switzerland and concluded with the convening of the second summit, held in Tunis between November 16 and 18, 2005. The Tunis Summit was preceded by four regional conferences, two subregional conferences, twenty-one WSIS thematic meetings and seven WSIS regional thematic meetings.[3] It culminated in the adoption of two political documents: the Tunis Commitment and the Tunis Agenda for the Information Society. The Tunis Summit, built upon the foundations of the Geneva phase, thereby reframed the WSIS's negotiation themes and converted previously adopted decisions into concrete initiatives.

Articulating the Two Phases of the Summit

Tunisia, host country for the second phase of the WSIS, strongly insisted that Phase II should be a high-level global event in its own right and that its political agenda should not be restricted to the narrow task of working out agreements on implementation and follow-up for the commitments made in Geneva. Tunisia's wish that Phase II be distinguished as a de facto stand-alone separate summit notwithstanding, the Tunis Phase was framed by the need to build on the gains, decisions, and commitments of Phase I. The only option was to conduct the second phase in the spirit of the first, and so the discussions at the Tunis Summit were strongly delineated by those of the Geneva Summit.

The general themes to be discussed at Phase II were defined in the Action Plan drawn up in Geneva. The Plan mandated the Tunis Phase to set up implementation and follow-up structures, reach consensus on the unfinished elements of the first phase, and draft documents that would reflect international consensus on the eradication of the digital divide.[4]

Various informal consultations took place early in 2004. However, the process did not get fully under way until the first preparatory committee meeting in June 2004. The mandate of this meeting was to achieve consensus on the political priorities for the second phase of the WSIS. At Phase II, Prep-Com I, follow-up and implementation of the Geneva Declaration of Principles and Action Plan, as well as the outputs of the Task Force on Financial Mecha-

---

3   As well as what has been described as "other WSIS-related meetings." See the WSIS Official Website at http://www.itu.int/wsis/preparatory2/index.html for full details on these meetings.

4   WSIS Executive Secretariat, *WSIS Plan of Action* (WSIS-03/GENEVA/DOC/5-E). (December 12, 2003). http://www.itu.int/wsis/docs/geneva/official/poa.html

nisms (TFFM) and the Working Group on Internet Governance (WGIG) were designated as topics for discussion for the Tunis Summit.[5]

What Was New, Different, and the Same at Phase II

Novelties and Changes in Phase II

Phase II of the WSIS had several characteristics that made it a unique international event in its own right.

From its outset, the agenda of the Tunis phase was shaped by its contrast with the preparatory phase of Phase I, during which the main problem facing governments had been the selection of suitable themes of discussion for the WSIS from among the myriad of possibilities. The goal of Phase I was indeed to formulate a common vision for the information society; this assumed it would be possible to achieve an international political consensus on the issues facing the WSIS. A considerable amount of time and resources had been dedicated to selecting discussion themes for the political agenda in Geneva. In contrast, the second phase was much more narrowly defined because it was limited to only a few themes, all of which had already been discussed in Phase I but had not been followed up by political commitments or implementation mechanisms.

Phase II of the WSIS emphasized the need for shared responsibility in the pursuit of the social, political, cultural, and economic goals defined in both phases. Responsibilities were divided accordingly among the many actors present, namely civil society organizations, members of the private sector, and international and governmental institutions, all of which were given different mandates. Indeed, one of the telling characteristics of the multi-stakeholder aspect of the WSIS was the principle of shared responsibility regarding the implementation, follow-up, and perfection of mechanisms that would concretize the WSIS outcomes. This raised the status of non-state actors to a level that politically confirmed their roles and responsibilities within the information society. Such esteem led several civil society members to express condemnation over what they perceived to be a lack of offical recognition in the task and responsibility allocation related to the eradication of information society inequality.

This acknowledgment, together with the commitment towards a multi-stakeholder system of governance, manifested itself in the rhetoric—and up to

---

5   WSIS Executive Secretariat, *Decision of PrepCom-1* (WSIS-II/PC-1/DOC/5). (June 26, 2004). http://www.itu.int/wsis/docs2/pc1/doc5.doc

a certain level in the work structures—of the major working groups formed during Phase II in the aim of guiding government negotiations. Working groups under the patronage of the UN Secretary-General were an innovation from Phase I. These working groups were both an opportunity for non-state actors to increase their participation in the summit, as well as strategic sites from which the political negotiations could be strongly influenced. While the degree to which the official rhetoric reflected the multi-stakeholder aspect of these groups can and should be questioned, the multi-stakeholder principle was clearly manifested in the working structure of the summit as the event took shape at the crossroads of diverging views on global governance.

A large portion of civil society efforts and resources that were devoted towards the WSIS during Phase I were aimed at promoting civil society's own active participation in the summit. The regulations and procedures for the participation of non-state actors, established during the PrepComs as well as at the summit itself, legitimized the presence of civil society at the WSIS while also limiting its participation in the decision-making process. Because these regulations and procedures were extended to Phase II of the WSIS, all non-state actors, whether they belonged to the private sector or civil society, could expect a base level of participation. Their goal was therefore to bolster this level of participation while avoiding any negative setbacks.

As soon as the participation guidelines for the major events of Phase II came to be known, civil society organizations rushed to expand their participation in more informal activities as well as in the new structures set up for the new phase. The majority of their efforts were therefore centered on their participation in the new working groups. Generally speaking, members of civil society effectively used the documents drafted in Geneva to reaffirm the multi-stakeholder principle of the summit while sidestepping its deficiencies.

A high-level international event held in the global South, far from conventional diplomatic hubs such as Geneva and New York, was a meaningful occurrence in and of itself. Although the fact that a UN summit was being held in Africa added an important symbolic component to the event, it also caused certain problems for the second phase. The most significant of these related to the prevailing political climate in Tunisia, which was marked by new participants acting directly or implicitly on behalf of the Tunisian government, whose ability to influence the WSIS represented a threat to principles of freedom of expression, of the press, and of opinion—the fundamental norms of a free and democratic information society. The efforts at agitation were effective, to a point. Throughout the preparatory process and continuing through the summit itself, the actions of the Tunisian regime aimed at silencing civil society and critical journalists actually functioned to attract a great deal of attention from media and the international community to issues of freedom of expression, information, and association (not to mention the government of Tunisia's lack of respect for each). Additionally, the summit being held in the South encouraged the emergence and recognition of new actors at the WSIS, which raised significant political issues for civil society.

In Continuity with Phase I

Official rhetoric for the WSIS centered around the notion of development, as summit organizers aimed to turn information and communication technologies into engines of social and economic development that would contribute to the UN Millennium Development Goals. This discourse did not significantly change during Phase II. However, much of the political tug-of-war that took place at the WSIS Phase II had more to do with geopolitics than it did with development. The reluctance of states to address the issue of information society financing compounded this lack of institutionalized will to turn the WSIS into an engine for the redistribution of communication resources. Nevertheless, Phase II of the WSIS was officially based on discourses of development, international solidarity, and social, economic and cultural inclusion. During both phases of the summit, the elimination of the digital divide remained a central topic for political discussions.

The multi-stakeholder nature of the summit was first established during the preparatory phase for Geneva, then confirmed by the Declaration of Principles and the Action Plan, and finally readopted and reaffirmed during Phase II. The multi-stakeholder approach marked the uniqueness of the WSIS in relation to other UN summits and was reiterated throughout the summit proceedings until their closing at Tunis. In addition, the multi-stakeholder approach was reflected in the follow-up and implementation mechanisms created from the decisions taken at the summit as well as in the decision-making bodies created or mandated in Tunis. Regardless, there remains a need for vigilance toward and critique of the official discourses of inclusion and transparency declared at the WSIS. Civil society remained deeply critical regarding the level and quality of the integration of these discourses into on-the-ground summit practice throughout the political proceedings.

The organizational structure of Phase II of the WSIS was very similar to that of Phase I. Following standard UN practice, the PrepComs were the main negotiation sites, receiving input from regional and thematic conferences associated with the WSIS. The rules and regulations for participation adopted during Phase I were readopted for Phase II, thereby avoiding a repetition of the tedious negotiations that characterized the preparatory process of the earlier phase. In addition, a so-called "Group of Friends of the Chair" was created and charged with supporting the work of the PrepComs by preparing resource documents. A relatively common structure within UN negotiations, it was hoped that use of a GFC in the second phase of the WSIS would help focus the debate in Phase II and avoid the extensive discussions of terminology and technicalities that had characterized Phase I.

Overview of Themes and Issues Discussed at WSIS

The documents agreed to at the conclusion of the Geneva phase reflected a consensus amongst governments on a number of often-broad ICT related themes.[6] But a series of important areas of contention remained. Issues surrounding Internet governance, information society financing, and the implementation and follow-up of the agreements made at the WSIS remained contentious until the very end of Phase I. Negotiations were therefore delegated to the Tunis phase in order to avoid the specter of political failure in Geneva. If it could reach a consensus on these complex issues, the Tunis phase would succeed where the Geneva phase had stalled.

In fact, the very concept of the "information society" around which the summit was organized remained contested by civil society and certain international organizations throughout both phases. In a joint declaration adopted at the end of the Geneva phase, civil society thought it necessary to state that:

> There is no single information, communication or knowledge society: there are, at the local, national and global levels, possible future societies; moreover, considering communication is a critical aspect of any information society, we use [...] the phrase "information and communication societies."[7]

---

[6] Including: the role of governments and all stakeholders in the promotion of ICTs for development; information and communication infrastructure as an essential foundation for the information society; access to information and knowledge; capacity building; building confidence and security in the use of ICTs; enabling environment; ICT applications as benefits in all aspects of life; cultural diversity and identity, linguistic diversity and local content; media; ethical dimensions of the information society; and international and regional cooperation. See the *Geneva Plan of Action* and *Geneva Declaration of Principles* at the WSIS Official Website: http://www.itu.int/wsis/documents/doc_multi.asp?lang=en&id=1161|1160

[7] WSIS Civil Society Plenary, *Shaping Information Societies for Human Needs: Civil Society Declaration to the World Summit on the Information Society*. (December 8, 2003). http://www.worldSummit2005.de/download_en/WSIS-CS-Dec-25-2-04-en.rtf

Civil society organizations thus avoided focusing strictly on the notion of "information" when contextualizing exchanges in the social and cultural realms of human communication. They also rejected the political, ideological, and technological implications of a single information society for the entire world, characterized by easy and instant access to information. Instead, civil society organizations chose to emphasize the local and complex nature as well as the cultural, social, and economic uniqueness of the human communities meeting over information networks.

In the meantime, international organizations, as well as governments, exhibited a range of approaches to the central themes of the WSIS. UNESCO, for example, also challenged the WSIS emphasis on information by developing its own rhetoric around the notion of knowledge:

> Knowledge societies are about capabilities to identify, produce, process, transform, disseminate and use information to build and apply knowledge for human development. They require an empowering social vision that encompasses plurality, inclusion, solidarity and participation. As emphasized by UNESCO during the first phase of the World Summit on the Information Society (WSIS), the concept of knowledge societies is more all-embracing and more conducive to empowerment than the concept of technology and connectivity, which often dominates debates on the information society....In other words, the global information society is meaningful only if it favours the development of knowledge societies and sets itself the goal of "tending towards human development based on human rights".... For UNESCO, the construction of knowledge societies "opens the way to humanization of the process of globalization."[8]

The WSIS political agenda had two closely associated objectives: eliminating the digital divide and harnessing the opportunities created by new technologies for the achievement of the Millennium Development Goals. The digital divide, which reflects the information and communication disparities existing between different populations and social groups, exacerbates political, social, economic, and cultural divisions in a world already divided. These divisions have become apparent in the development and distribution process of the technological tools necessary for contemporary communication. The digital divide can thus be said to be an echo or an incarnation of a series of preexisting divisions manifested in the access to and use of technology.

The digital divide is in fact a phenomenon that conceals a much larger communication barrier. Issues stemming from access to and acquisition of both contemporary and traditional information and communication technology conceal crucial matters of access, knowledge-sharing, knowledge itself, and

---

8   UNESCO, *Towards Knowledge Societies: UNESCO World Report.* (2005). unesdoc.unesco.org/images/0014/001418/141843e.pdf

culture. Indeed, the digital divide can in no way be considered separately from its underlying basis of literacy, access to electricity, and the availability of basic phone service and television and radio broadcasting systems. Without doubt, an information society requires a basic level of education for all. According to UNESCO, more than 774 million adults worldwide are illiterate, the majority of them women.[9] Problems around bringing electricity to isolated communities persist, and considerable progress remains to be made for providing vulnerable populations with access to basic education. All things considered, the digital divide illustrates a set of problems related to structural inequality more so than the proliferation of new modes of information production, diffusion, and consumption.

That said, the disparities within various populations in terms of their access to and ability to make meaningful use of ICTs were stark and troubling by the beginning of the WSIS. For example, ITU data presented in a report published in 2004 revealed that, in the period around the WSIS first phase:

- Less than 3 out of every 100 Africans used the Internet, compared with an average of 1 out of every 2 inhabitants of the G8 countries (Canada, France, Germany, Italy, Japan, Russia, the UK and the US).
- There were roughly around the same total number of Internet users in the G8 countries as in the whole rest of the world combined:
  - 429 million Internet users in G8
  - 444 million Internet users in non-G8.
- The G8 countries were home to just 15% of the world's population—but almost 50% of the world's total Internet users.
- The top 20 countries in terms of Internet bandwidth were estimated to be home to roughly 80% of all Internet users worldwide.
- There were more than 8 times as many Internet users in the US than on the entire African continent.
- The entire African continent—home to over 50 countries—had fewer Internet users than France alone.
- There were still 30 countries with an Internet penetration of less than 1%.
- The 14% of the world's population living in the G8 countries accounted for 34% of the world's total mobile users.
- Of Africa's 26 million fixed lines, over 75% were found in just 6 of the 55 African nations.
- Africa had an average of 3 fixed lines per 100 people.
- The Americas region had an average of 34 fixed lines per 100 people.

---

9   UNESCO Institute for Statistics, *According to the Most Recent UIS Data, There Are an Estimated 774 Million Illiterate Adults in the World, About 64% of Whom Are Women.* (Updated on October 8, 2009). http://www.uis.unesco.org/ev_en.php? ID=6401_201& ID2=DO_TOPIC

- The European region had an average of 40 fixed lines per 100 people.[10]

While these statistics are now clearly outdated, they provide an effective snapshot of the situation at the international level at the beginning of Phase II and contextualize the steps taken at the WSIS towards reducing the digital divide.

Unresolved Issues of Phase I

Internet Governance

Internet governance remained a profoundly controversial topic throughout both phases of the WSIS. A confrontational and oppositional dynamic prevailed and lasted until the end of the summit proceedings, after which it was transferred to the political bodies that resulted from the WSIS.[11]

The political negotiation process first sought to agree on a shared definition of Internet governance. This preliminary process remained instrumental in shaping negotiations around the issue, as the agreed to definition would identify (and thereby demarcate) any technical and political problems that would need to be addressed by political intervention.

Two visions of Internet governance emerged over the course of the WSIS. The first approach, vigorously defended by government delegations from Organization for Economic Co-operation and Development (OECD) countries, the international private sector lobby and a number of groups representing Internet engineers and technologists, defined Internet governance as a mostly technical process involving the coordination and allocation of system resources to ensure network efficiency, fluidity, and adaptability. This perspective marginalized political issues linked to the organization of the system and essentially framed Internet governance as contingent on administrative and technical processes.

A second definition of Internet governance, this one presented by a coalition of developing countries and regional powers, included the more political dimensions of accessibility, transparency, and political control of the network. This perspective sought to politicize the resource coordination and allocation process that is at the heart of the Internet structure by introducing highly controversial issues into the discussions on Internet governance. This involved

---

10 These statistics are available online and are a rendition of the evaluation made by the ITU in 2004. See *What's the State of ICT Access Around the World?* At http://www.itu.int/wsis/tunis/newsroom/stats/

11 The Internet Governance Forum, one of the more significant legacies of the WSIS, will be analyzed in detail in Chapter 7.

questioning the roles and positions of different states vis-à-vis the management of the Internet. States were therefore faced with three general options for Internet governance: maintaining the status quo and affirming the efficiency and legitimacy of the system currently in place, modifying that system either slightly or drastically, or completely reshaping the Internet governance system by redefining the roles and responsibilities of the parties involved in the governance process.

Political negotiations reached an impasse during Phase I of the WSIS when no consensus on an appropriate definition for Internet governance could be reached. Summit actors avoided complete diplomatic failure by using a technique that has been tried and tested in the field of international relations: the negotiations were postponed to a later date. The Tunis Summit became the cut-off date for reaching a consensus on the issue.

There was, however, a general acknowledgment in Geneva that a Working Group should be created and given the responsibility of defining, describing, and presenting the options available to governments. The work of the Working Group would thus support and inform the political negotiations on Internet governance. The mandate of the Working Group in regards to Internet governance was defined in the Action Plan devised in Geneva. Following the Plan, the UN Secretary-General was asked:

> to set up a working group on Internet governance, in an open and inclusive process that ensures a mechanism for the full and active participation of governments, the private sector and civil society from both developing and developed countries, involving relevant intergovernmental and international organizations and forums, to investigate and make proposals for action, as appropriate, on the governance of Internet by 2005.

For members of civil society mobilized around the WSIS, Phase I therefore ended with high expectations for active participation in a working group that would strongly contribute to establishing international policies on communication.

We will discuss the role of this working group as it pertains to the structure of the debates on Internet governance in detail in Chapter 5. However, it is worth mentioning that, due to the sizeable participation of civil society in this multi-stakeholder working group (nearly a third of the members of the working group came from different civil society organizations), and despite deep internal controversy around the selection of these civil society members, the working group was hailed as an important victory at the WSIS.

Financing the Information Society and International Cooperation

The WSIS took place in a context marked by the maintenance and even reinforcement of deep social and economic inequalities at the international level. Summit organizers were conscious of the exclusion of entire segments of different populations from the information society and of the challenges such access gaps create for the economic development of entire regions. Many summit actors came to see the information society financing issue as central to the eradication of poverty, because this financing would serve as international aid that would integrate these populations into global communication networks. Although development issues were not reduced solely to financing debates, the role of financing became one of the most discussed issues related to development.

A number of states led by Senegal insisted that the development of new information society financing mechanisms should become a WSIS priority. This stand was marginalized by a number of developed countries. Those in favour of new financial transfer mechanisms ran up against the weight of such powers as the European Union and the United States, who preferred to see international cooperation and digital solidarity develop through existing mechanisms and institutions. Developed countries were not willing to undertake new expenses or increase their existing financial contributions, creating a political context in which it was difficult to advocate for the principle of a more equitable distribution of existing resources

This disagreement endured the entire first phase; no consensus could be reached at Geneva. As a result, summit actors decided to mandate a working group to evaluate the issues and present them with proposals for action during the second phase, just as they had done with the issues surrounding Internet governance.

Information society financing was included in official documents of the summit under the heading of a "Digital Solidarity Agenda." This Agenda "aimed at ensuring that all conditions for the mobilization of human, financial, and technological resources required for all men and women to participate in the new information society are met."[12] The Declaration of Principles adopted in Geneva effectively supported the Agenda.

> 17. We recognize that building an inclusive Information Society requires new forms of solidarity, partnership and cooperation among governments and other stakeholders, i.e. the private sector, civil society and international organizations. Realizing that the ambitious goal of this Declaration—bridging the digital divide and ensuring

---

12 WSIS Executive Secretariat, *Geneva Plan of Action* (S-03/GENEVA/DOC/5-E). (December 12, 2003). Section D27. http://www.itu.int/wsis/docs/geneva/official/poa.html

harmonious, fair and equitable development for all—will require strong commitment by all stakeholders, we call for digital solidarity, both at national and international levels.[13]

The Geneva Plan of Action was to concretize the principle of solidarity adopted in the Declaration of Principles by means of political and administrative initiatives that would effectively reduce inequality in access and use of information and communication technologies. The Plan called for the analysis of the financial mechanisms already in place at the end of the first phase. Heads of state taking part in the Geneva phase could thus avoid embarking on new projects and amassing new expenses while publicly announcing their intent to act on the financing issue. The Action Plan was therefore severely criticized by civil society groups taking part in the WSIS, who saw it as lacking in concrete commitments on the part of governments for the achievement of the goals outlined in the Declaration of Principles, promoting an unsuitable ideology (the Plan of Action made important mention of free-market development and creating an environment favourable for the growth of commerce), as well as lacking in vision. Civil society organizations were equally critical of the cold responses that greeted the calls for the establishment of new financial mechanisms such as a digital solidarity fund. These conservative responses came to considerably restrain the mandate of the Task Force on Financial Mechanisms (TFFM) created at the end of Phase I, which was ultimately limited to evaluating existing financial mechanisms.

Stocktaking and Implementation

Both summit organizers and the Tunisian government officially presented the Tunis Summit as a "summit of solutions." By the end of Phase I, no concrete plan to transform the Action Plan into effective implementation and follow-up measures had been formulated. The eve of Phase II was therefore spent redistributing responsibilities to the different WSIS actors, determining follow-up and implementation policies, and defining metrics by which to evaluate actions taken and results.

Despite the fundamental importance of the question of how decisions taken at the WSIS would be implemented in practice, evaluated and followed-up on, the issue would remain largely overshadowed for much of Phase II as political and media attention around the summit focused on more sensational issues such as Internet governance and information society financing. As the summit approached its conclusion at PrepCom III however, implementation

---

13 WSIS Executive Secretariat, *Geneva Declaration of Principles* (WSIS-03/GENEVA/DOC/4-E). (December 12, 2003). http://www.itu.int/wsis/docs/geneva/official/dop.html

and follow-up mechanisms emerged as the topic of a focused, controversial and important debate in their own right.

Follow-up and evaluation was outlined in section 28 of the Geneva Plan of Action, and the mandate of Phase II in regard to this was described in section 29 of the same document:

> A realistic international performance evaluation and benchmarking (both qualitative and quantitative), through comparable statistical indicators and research results, should be developed to follow up the implementation of the objectives, goals and targets in the Plan of Action, taking into account different national circumstances.
>
> [...]
>
> Follow-up and implementation of the Geneva Plan of Action at national, regional and international levels, including the United Nations system, as part of an integrated and coordinated approach, calling upon the participation of all relevant stakeholders. This should take place, inter alia, through partnerships among stakeholders.[14]

This blueprint was to serve as a basis for the fundamental issues discussed in Phase II. The "multi-stakeholder" aspect of the issues and their resolutions were emphasized repeatedly in the official documents of the summit. As such, civil society organizations present at the WSIS would significantly contribute to the critical analysis of these issues as part of official summit proceedings; nevertheless, they would first have to organize themselves according to both the ethical principles they claimed to uphold and the political and institutional constraints of the summit.

Themes Raised by Civil Society in Phase II

Social Justice, Financing, and People-centered Development

The WSIS was very much in line with other UN events that aimed to contribute to the achievement of the Millennium Development Goals.[15] Following the summit, information and communication technology was to be put to use toward the elimination of famine, disease, misery, and poverty, while reinforcing democratic principles and institutions.

---

14 WSIS Executive Secretariat, *Geneva Plan of Action* (WSIS-03/GENEVA/DOC/5-E). (December 12, 2003). http://www.itu.int/wsis/docs/geneva/official/poa.html

15 See, for example, the Earth Summit of 2002 (http://www.earthSummit2002.org/) and the 2005 World Summit (http://www.un.org/Summit2005/). See also the Millennium Development Goals website for more information on specific events and initiatives. http://www.unhabitat.org/categories.asp?catid=535

The WSIS recognized the importance of connecting all countries to information, knowledge, and communication networks in order for these benefits to reach all of the vulnerable groups, nations, and regions faced with different forms of social and economic exclusion. Nevertheless, due to endemic poverty, many nations could not afford to develop the infrastructure, train the workforce, and integrate and maintain the technology required for such a project. As a result, financial, technological, and knowledge transfer mechanisms became a priority for the summit, which closely linked information and communication technology to international development.

The summit therefore strove to guide the international community to an agreement on a definition for the information society and the issues at hand (especially development issues), as well as to bring concrete decisions to the more urgent problems. The issue of financing became problematic from the outset: current mechanisms for cooperation in international development remained ineffective, and developed countries systematically refused to heed demands for the redistribution of wealth and adoption of restrictive agreements on financing.

A Digital Solidarity Fund that called for voluntary contributions from cities and states was set up at the initiative of the government of Senegal in the period leading up to the Geneva Summit. While the initiative did not gain official support from the government delegations in Geneva, the Digital Solidarity Fund was lauded by many civil society organizations at the summit.

Civil society organizations upheld several interesting positions on the financing of the information society for development purposes. In particular, different organizations conceptualized information and communication as a global public good. Even though infrastructure, networks, and technologies can be and often are monopolized by private interests, the use of the content and information passing through these media by one individual does not diminish their utility for anyone else. On the contrary, having more people connected to the network and new knowledge and information pass through it only increases its utility for everyone. Several groups presented information and knowledge as public goods in this way.[16] The information-as-public-good approach quite obviously came up against its polar opposite, an approach that advocated mostly privatized communication networks. The tension between the two partially explains why many civil society organizations placed so much emphasis on the public financing of infrastructure, the need to incorporate free software into development projects, and the promotion of community

---

16 See, for example, the work done during the WSIS and since by the Bangalore-based NGO IT for Change. http://www.itforchange.net/.

initiatives. According to this approach, intellectual property rights are strong factors of social, cultural, and economic exclusion that would ensure that certain populations continue to live illegally or in poverty.

Human Rights

A large portion of civil society activity during Phase I of the WSIS was devoted to the promotion of human rights. Civil society fought hard to include human rights discourses in official WSIS documents during the preparatory process leading up to Geneva. The results of these efforts were mixed. Although the Geneva Declaration placed the WSIS in line with notions of development and respect for human rights, it avoided mentioning any specific or overlying rules that would allow for the development of a democratic, open, and inclusive information society.

During Phase II, certain human rights issues deemed particularly important were repeatedly raised by civil society. These issues, namely relating to freedom of expression, opinion, information, and association, created a considerable amount of discussion both in the WSIS arena and among civil society organizations themselves. Initially, given Tunisia's regrettable performance in confronting issues around freedom of expression and information, several actors proposed boycotting the WSIS and demanding that it be relocated to a country with a better human rights record. For many CS actors, allowing a World Summit to take place in a country known for silencing the press and opponents of the regime was paradoxical. These actors furthermore expressed serious concern for the security of the international participatnts in the event.

Civil society organizations also emphasized the paramount importance of privacy concerns in a context of excessive cyber-surveillance, abusive and illegal electronic listening, and private data monitoring, transfer, and exchange without user consent. The global context within which the WSIS was organized, marked by the war on terror and powerful private actors asserting their control of financial and technological capital, reinforced civil society's position calling for the protection of the private lives of citizens.

The discussions of Internet governance during Phase I also encouraged participating civil society organizations to remind government delegations of the strong links between Internet governance and issues in freedom of expression, association, thought, religion, and the press. Civil society further insisted that a democratic Internet governance regime must ensure respect for private life and facilitate access to knowledge and culture.

By and large, civil society deplored the absence of mechanisms and procedures that would ensure the protection of and respect for rights and liberties in a democratic information society.[17]

Internet Governance

As we discuss in our long view of the participation of CS within the Internet governance debate presented in Chapter 7, the emergence of Internet governance as a primary political issue of Phase II instigated something of a changing of the guard within CS. Some of the prominent CS voices and organizers in Phase I had little interest in or experience with the geopolitics of Internet governance, many of those associated with the Communication Rights in the Information Society (CRIS) campaign in particular. As this first generation of WSIS CS stepped back, there was an influx of new CS actors who gravitated to the WSIS Phase II. Many of these new IG-focused CS participants arrived with relatively well-set opinions formed through years of previous experience participating in the Internet Corporation for Assigned Names and Numbers (ICANN) and other Internet policy forums. These views were often mutually exclusive and competing.

CS would find common ground on insisting that Internet governance be made more transparent, open and legitimate by whatever agreements were reached at Tunis and on generally expressing a strong preference for multi-stakeholder models in place of new governmental structures and existing governmental control. In particular, CS rallied behind the work of the WGIG—which bore the distinct influence of its many CS participants—and, in particular, the idea proposed in the WGIG report to create a multi-stakeholder forum for continuing the Internet governance discussion beyond Tunis.

Global Governance and Participation at the WSIS

Although not directly related to the general themes of the WSIS, global governance and multi-stakeholder participation quickly became one of the more significant issues of Phase II. While it was perhaps limited in terms of its substantive agenda, the WSIS presented a major opportunity for civil society to establish a legacy for its own participation in the global governance of communication. As a result, summit activity would therefore have to transform

---

17 WSIS Civil Society Plenary, *Much More Could Have Been Achieved: Civil Society Statement on the World Summit on the Information Society*. (December 18, 2005 Revision 1—December 23, 2005). http://www.worldSummit2003.de/download_en/WSIS-CS-Summit-statement-rev1-23-12-2005-en.pdf

new methods for consultation, inclusion, and participation enlargement into policies derived from the decisions that were taken at an international level.

The repeated effort on the part of civil society organizations to address the issue of participation at the WSIS illustrated their desire to be included in the decision-making process of the summit. In fact, the integration of new non-state actors into an international arena represented a policy issue in and of itself:

> The substantial and institutional nature of WSIS could be seen as an attempt to build a new model for global governance [...] WSIS is a test of the capacity of the multilateral system to find alternative and innovative ways to integrate a wider range of actors in a long-standing political process, in order to deal more adequately with the challenges raised by the Information Society.
>
> [...]
>
> The repeated reference to the multi-stakeholder nature of this Summit, as well as the development among civil society representatives of a critical awareness of their role, has therefore represented one of the most important precedents within the WSIS. There is a broad agreement that this approach will guarantee the success of the Summit.[18]

Civil society organizations taking part in the WSIS demonstrated their desire to promote permanent change regarding the way that the UN system incorporates civil society into its proceedings. To this end, a significant portion of civil society efforts went towards ensuring that official WSIS texts reflected a multi-stakeholder approach to communication governance and pushed to create a transition from a multilateral, intergovernmental mode of governance to a multi-stakeholder approach. These efforts were underpinned by discourses centering on notions of transparency, inclusion, legitimacy, democratic participation, and public responsibility. The claim being forwarded by CS was that the degree to which the WSIS decision-making processes were open to civil society organizations demonstrated the extent to which the WSIS could be said to reflect such principles.

Themes Raised by Civil Society in Both Phases

The following cross-cutting themes were also formulated during both the first and second phases of the WSIS:

---

18 Conference of Non-Governmental Organizations in Consultative Relationship with the United Nations (CONGO), *Civil Society Orientation Kit*. (November 2005). p.35 http://www.ngocongo.org/congo/files/wsis_oriention_kit.pdf

Fighting Exclusion: The Struggle for Inclusivity in the Information Society

The global information society currently taking shape excludes entire segments of the world's population on a daily basis. The causes for exclusion are numerous and shaped by cultural, social, linguistic, economic, and regional variables, as well as factors in gender, age, mobility of individuals, and ethnic composition of different social groups. The following issues represent the core of the rhetoric put forward by civil society during the two phases of the WSIS:

- gender issues: gender equality in regards to the access, use, and development of information and communication technology;
- linguistic marginalization: the exclusion of minority languages in networks and media, linguistic domination of a few languages in software tools;
- accessibility issues: problems regarding reduced mobility: elderly and handicapped people in the information society;
- equity issues: youth, women, social and ethnic minorities: inclusion and participation in the face of majority and/or dominating social groups;
- development issues: international solidarity and connecting Africa, Asia, and South America; regional characteristics, common issues and capacity building;
- education and training issues: provide individuals, groups, and communities with the means for developing local expertise based on local experiences in order to encourage inclusion and development.

Intellectual Property Rights, Patents, Trademarks, and Public Domain: Opening Cracks in the Walls of the Information Society

This cross-cutting theme touched on many aspects of an inclusive information society. Issues in development cannot be adequately addressed without taking into account the costs incurred by software and proprietary technology that are designed for the consumption of creditworthy segments of society. Civil society organizations therefore promoted alternative production, dissemination, and appropriation methods for digital technologies characterized by the sharing of knowledge, technical cooperation, and collaborative exchange and assistance. Several civil society organizations also used the WSIS to draw attention to broader issues in the international intellectual property rights regime.

The following topics were at the core of discussions dealing with issues related to intellectual property at the WSIS:

- available, free, and open software;
- the role of the state in promoting and protecting marginalized cultures and groups;

- reforming the international intellectual property regime;
- commercial agreements, trade liberalization, protection of national, regional, and local cultures.

Media Matters: It's Not All About ICTs

Civil society organizations fought hard throughout both WSIS phases to ensure that traditional media would not be forgotten at the summit. The written press, radio, and television continue to play a considerable role in the social, political, cultural, and economic development of marginalized populations. The social and political issues linked to the transnational capitalist political economy of information and communication were also brought to light by civil society at the summit.

The issues raised by civil society in regards to traditional media focused on the following:

- concentration of media and press ownership, distance between media and citizens, domination of media by large corporations;
- civil society participation in media governance and communication policy development;
- the media and education;
- the media and development;
- the media and gender;
- communication rights and access to media institutions;
- freedom of the press and freedom of expression;
- alternative and community media.

The often large gaps between government positions and those taken by civil society organizations on the issues presented in this chapter would eventually test the multi-stakeholder participation mechanisms adopted by the WSIS. The opportunities for civil society groups and organizations to adequately convey their positions to government delegations that were already involved in difficult political negotiations were unequal throughout the preparatory process for Phase II. In addition to dealing with the participation issues and content discussed above, debates in Phase II were also shaped by organizational issues affecting WSIS civil society itself.

Civil Society at WSIS: Backgrounds, Structures and Practices

Rules and Procedures for Civil Society Participation

The rules of stakeholder participation at the WSIS were established through a series of regulations, procedures, and practices that were negotiated by government delegations. The terms of participation for WSIS observers were primarily laid out in three formal documents adopted at the beginning of Phase I:

- Arrangements for Accreditation;[19]
- Arrangements for Participation;[20]
- Rules of Procedure of the Preparatory Committee of WSIS.[21]

The first of these documents, Arrangements for Accreditation, described and defined the procedures that organizations wishing to participate in WSIS would have to follow. It also outlined the information that organizations seeking accreditation would be required to present to summit organizers in order to fulfill the selection criteria for prospective participants. Of particular importance were details regarding organizations' activities, financing, and membership. This accreditation process ensured that any organization accepted to participate in any single WSIS event would also be accredited for participation in all other summit events.

The second document, Arrangements for Participation, focused on observer participation in WSIS preparatory events. The document granted civil society and private sector entities with observer status, giving them the right to participate in the summit's preparatory process. Furthermore, it described the right of non-state actors to submit written contributions, to be distributed by the Executive Secretariat, as well as the ability to nominate spokespeople to make interventions on behalf of stakeholder groups.[22]

These modalities of participation were also specified in sections 51 to 57 of the *Rules of Procedure of the Preparatory Committee*. The privileges, rights, and

---

19 WSIS Executive Secretariat, *Arrangements for Accreditation Adopted at the First Session of the Preparatory Committee*. (July 1-5, 2002). http://www.itu.int/wsis/docs/pc1/official/arrangements_accreditation.doc
20 Ibid.
21 WSIS Executive Secretariat, *Rules of Procedure of the Preparatory Committee Adopted at the First Session of the Preparatory Committee*. (July 1-5, 2002). http://www.itu.int/wsis/docs/pc1/official/rules_procedure_pc.doc
22 Ibid.

measures applicable to observers differed according to the category to which the observer belonged. The following observer categories were established for the WSIS:

- entities and organizations having received a standing invitation to participate as observers in the sessions and work of the General Assembly;
- UN Secretariat and organs (including UN funds and programs);
- UN specialized agencies;
- other invited intergovernmental organizations;
- accredited civil society entities (including NGOs in consultative status with ECOSOC);
- accredited business sector entities (including ITU Sector Members);
- associate Members of Regional Commissions.[23]

In this way, official WSIS regulations allowed observers access to preparatory and subcommittee meetings and established the nature of the dynamics between these observers and government actors. Access to these meetings became more and more crucial as the work progressed. The regulations remained noticeably vague in regard to several key issues; in practice, stakeholder participation issues were often treated on a case-by-case basis during negotiations. Certain summit practices, despite not being formally described in the observer participation regulations, were adopted as vested rights of the organizations taking part in the WSIS. As a result, stakeholder participation rights were often granted through subjectively liberal interpretations of the rules by summit organizers (in particular, the PrepCom presidents, Adama Sammasékou of Mali in Phase I and Janis Karklins of Latvia in Phase II). But, absent clearer catch-all regulations, during tense negotiations that urgently required a political consensus on difficult issues, questions about stakeholder participation seemed to be strategically raised by certain heel-dragging governments who were not interested in seeing the status quo effectively challenged, and the rights of stakeholders to participate in negotiation groups were often contested by certain governments and even withheld when representatives of such governments were charged with chairing such sessions.

Civil society organizations capitalized on the progress toward greater inclusion of non-state actors that had been made during Phase I by successfully gaining access to intergovernmental subcommittee meetings. Drafting groups were also formed by subcommittees in order to negotiate the precise language

---

23 WSIS official website, *Basic Information about WSIS: The Multi-stakeholder Participation in WSIS and Its Written and Unwritten Rules.* http://www.itu.int/wsis/basic/multistakeholder.html

that would appear in the final agreed texts. WSIS official regulations did not address the participation rights of stakeholders in these drafting groups, and this emerged as a highly sensitive issue and one that was crucial to civil society's full participation in the summit's political process. Exclusion from the drafting groups would have denied non-governmental stakeholders access to up-to-date information about the negotiations and seriously limited opportunities for lobbying. More fundamentally, closing the door to civil society observers at the crucial moment when governmental delegations effectively put pen to paper would have limited the ability of civil society to hold governments accountable to the more general statements made in plenary meetings at the moment when they were required to propose and/or support or critique precise text for WSIS outcome documents. Establishing more or less unofficial conventions to keep all meetings associated with the WSIS process open to observer participation therefore became a considerably important strategic issue for civil society and for the private sector as the gains toward multi-stakeholder governance made in principle during Phase I effectively eroded in practice as political negotiations became more complex over the course of Phase II.

Implementing the Multi-Stakeholder Global Governance Model

Through negotiation with government delegations, summit organizers designated several different participant categories for observers at the WSIS and six different tracks of participation were set up for stakeholders at the WSIS:

1. submit written contributions;
2. make verbal comments;
3. assist in the working groups and committees;
4. intervene during official proceedings;
5. participate in discussions with government representatives in round table meetings and panel discussions;
6. convene in WSIS-designated meeting sites.

Among the many national delegations present at the summit, the degree of enthusiasm for allowing observer participation varied greatly from one country and one region to another. Certain countries, notably China, Iran, Egypt, and Cuba, were particularly averse to granting extended participation rights to civil society organizations and other observers in the WSIS. Others, such as Germany, Canada, and the European Union, took on a more inclusive stance. Participatory practices remained inconsistent throughout the summit, however, and, at times, certain governments seemed to capitalize on this uncertainty to push for ad hoc decisions that either enhanced or restricted CS participation in specific meetings or processes. Thus, the flexibility of the summit, which contributed greatly to developing its multi-stakeholder aspect, also served as a strategic tool of exclusion and inclusion during periods of crucial negotiations.

Civil Society Structures at the WSIS

Five structures were put into place during the WSIS in order to promote, frame, and encourage civil society participation in the debates and activities held at the event. Two of the five structures were officially recognized by both the summit organizers and government delegates. The other three structures remained endogenous to civil society and anchored its autonomous organization at the WSIS.

Civil Society Official Structures

Civil Society Division (CSD)

The first phase of the WSIS saw the creation of two major official structures meant to encourage and facilitate the participation of civil society at the WSIS. The first of these structures, the Civil Society Division, was one of four divisions of the WSIS Executive Secretariat, the body in charge of planning and organizing the summit. The other three divisions focused on governments, the private sector, and intergovernmental agencies. These divisions were intended to reflect the multi-stakeholder aspirations of the WSIS and ensure integration, participation, and coordination of the different actors invited to participate in it. The mandate of the Civil Society Division was to:

- brief all actors on events and information pertinent to the summit;
- provide civil society participants with the information and working materials necessary for their full inclusion in the preparatory process;
- inform other summit participants of civil society's concerns;
- facilitate workshops and seminars on key issues affecting civil society;
- guide on-line discussion groups of civil society participants;
- work closely with the media to ensure that the issues of civil society were heard;
- collaborate with other divisions of the Executive Secretariat;
- seek new perspectives on the issues relevant to the summit's agenda.[24]

---

24 See Marc Raboy and Normand Landry, *Civil Society, Communication and Global Governance: Issues form the World Summit on the Information Society*. New York: Peter Lang, 2005, p. 50.

The Civil Society Division was established at the end of 2001 and continued operations through the entire first phase. As financing for the CS Division ended with the closing of the first phase and no other sources of financing could be found, the Civil Society Division was disbanded in early 2004. This constituted a major setback that sent a negative message to civil society participants about the prospects for civil society inclusion in the Tunis phase.

Civil Society Bureau (CSB)

Set up during the preparatory process of Phase I, the Civil Society Bureau was created in response to the massive presence of civil society and its endogenous structures (described below). Government delegations approved the establishment of a Civil Society Bureau to work with the Secretariat and its Intergovernmental Bureau at the second PrepCom of the first phase of the WSIS. For the first time in the history of the United Nations, a civil society structure was officially recognized by participating governments, alongside the existing governmental structures for facilitating civil society participation. In contrast to the Civil Society Division, which was composed of Executive Secretariat employees, members of the Civil Society Bureau were chosen from the various organizations participating in the summit.

During Phase I of the WSIS, the Civil Society Division ensured liaison between the Civil Society Bureau and the Executive Secretariat. During Phase II, this role was essentially transferred to the Conference of Non-Governmental Organizations in Consultative Relationship with the United Nations (CONGO), a non-governmental organization with a professional staff that is deeply integrated with the UN system.

The Civil Society Bureau was mandated with a mainly logistical role: facilitate exchanges between civil society and government representatives, improve communication, contribute to the organization of civil society participation in WSIS events, and serve as a conduit for procedural matters involving multiple WSIS decision-making bodies. From the outset it was agreed that the Bureau should concern itself solely with logistical issues and would not prepare or submit content that could in any way be interpreted as an effort to influence the political process.

By the end of the second phase, the Civil Society Bureau was composed of 22 "families" grouping together several different organizations with similar interests, positions, or origins.[25] One individual in each of these families

---

25 These included "thematic families" (Cities and Local Authorities; Education, Academia and Research; Finance; Gender; Indigenous People; Internet Governance; Networks and

served as a "focal point" for communicating information regarding new developments in the WSIS and the Bureau to family members.

Endogenous Structures

Civil society, meanwhile, had created its own endogenous structures to facilitate participation in and internal organization at the WSIS. The three most important were the Civil Society Plenary, the Content and Themes Group (CS C&T) and the series of civil society caucuses and working groups. Online discussion lists for each of these structures were also set up.

During both WSIS phases, civil society was coordinated in a multi-level participatory process that involved online and in-person meetings of these actors. Caucuses and working groups exchanged ideas on electronic email lists. This communication tool became indispensable not only for discussions on content themes, but also for the internal management of activities, events, and the work of the many groups that were mobilized around the WSIS. Thematic and organizational consultations took place both on endogenous lists and via the Civil Society Bureau. The Content and Themes list encouraged debates and exchanges on WSIS issues and fueled the drafting process for positions related to them. In March 2005, the list had around 230 members. The CSB list allowed members of the Bureau to exchange ideas on issues related to the organization and work of the Bureau. Finally, the Civil Society Plenary list encouraged transparency and debate between members of civil society wishing to address large strategic, organizational, and decision-making issues related to civil society participation in the WSIS. Around 500 members were registered to the plenary list, which remained strictly for discussion and did not aim to lead to decisions. A series of additional email lists with more specific and indirect mandates linked to the WSIS were also established over the course of the process.

Civil Society Plenary (CSP)

The Civil Society Plenary was created autonomously by individual and group delegates participating at the first PrepCom of Phase I in July 2002. The Plenary immediately became civil society's main decision-making and deliberative body at the summit and remained so until the summit's conclusion. It con-

---

Coalitions; Media; Multi-stakeholder partnerships; NGOs; People with Disabilities; Philanthropic Institutions; Science and technology commuity; Trade Unions; Volunteers; Youth) and "regional families" (Africa; Asia-Pacific; Europe and North America; Latin America; Western Asia and the Middle East; Host Country Liaison). See WSIS Civil Society Bureau Website. http://www.csbureau.info/contactinformation.htm

vened at all official WSIS meetings attended by civil society, and all civil society delegates were free to join and participate in its work. The Plenary considered issues of common interest to civil society as a constituency, including those related to summit participation, organization, and agenda-setting. Plenary meetings were convened to deliberate on such issues and determine if a CS consensus view was emerging. During the second PrepCom of Phase I, the Plenary endorsed the creation of the Content and Themes Group in addition to the already existing Civil Society Bureau, thereby effectively legitimizing these structures. The online component of the Civil Society Plenary, managed by the Virtual CS Plenary group (plenary@wsis-cs.org), included a discussion list that was intended to help facilitate an ongoing exchange of ideas between CS delegates, enable some form of remote participation for those who could not attend all WSIS events in person, and keep civil society actors informed about developments occurring between the main WSIS meetings. Although the Virtual CS Plenary group was in no way intended to be a decision-making body, the list remained the principal strategic meeting place for discussing and organizing civil society activity online throughout WSIS proceedings. The list remained open to all civil society organizations and accredited WSIS groups as well as anyone who may not have received official status at the summit but demonstrated interest in one or more of its major themes.

The role of the CSP changed noticeably from Phase I to Phase II. Its role as a preference-measuring and decision-making body declined during the Tunis phase as deeply held differences of opinion emerged within civil society that effectively eliminated the prospects for arriving at a civil society consensus view on much of the WSIS agenda.

Civil Society Content and Themes Group (CS C&T)

The Civil Society Content and Themes Group was created in July 2002. Its mandate was to coordinate the work of the different caucuses and working groups, contribute to consensus-building on substantive issues of common interest, and facilitate the drafting of common positions to be transmitted to WSIS organizers and participants. The group's work included coordinating production, drafting and revision of collaborative documents, working out civil society speaker nominations, and translating documents. Most notably, C & T coordinated the drafting and adoption of joint declarations and position statements on various issues representing consensus views of civil society that constituted official summit documents and statements of civil society positions on the content and issues related to official intergovernmental negotiations.

The role of the CS C&T was influential throughout the drafting process for the final civil society declaration of Phase I. By contrast, following the closing of Phase I, the group's coordinators withdrew, new participants emerged, and a general reevaluation and reorganization of civil society structures was undertaken. The sum result was that the momentum and dynamism developed by C & T over the final stages of Phase I had largely dissipated over the course of the transition to Phase II.

Caucuses and Working Groups

Caucuses and working groups were at the centre of the bottom-up activities that best characterized civil society participation at the summit. Clusters of organizations and individual CS delegates with common interests were initially encouraged to create or merge into working groups or caucuses to be organized on some combination of interest, position, experience, or background proximity.[26] Creation of these caucuses and working groups helped to organize CS so that expertise could be concentrated and networked in the aim of producing high-quality contributions on specific, clearly defined themes and issues. The work of the caucuses and working groups supported both the official proceedings and the more informal discussions and brainstorming that were internal to civil society at the WSIS.

The loose network structure of the caucuses meant that various groups were also able to work collaboratively with each other, to discuss issues of common or overlapping concern, draft joint positions, and formulate joint strategies. Whenever a group or caucus reached a consensus on a given theme, it could send its contributions to CS C&T to be integrated into larger documents, or even send them directly to the WSIS secretariat to serve as background information contributed to the official proceedings. Despite the creation of this self-consciously flat organizational structure, certain caucuses emerged as undeniable focal points for civil society participation while others

---

26 These included "strategy, coordination and logistic caucuses and working groups" (CS Plenary; CS Content and Themes; General list; Coordination list; Drafting team; Speakers nomination Committee; Translation team; CS Bureau; WorkingMethods WG); "regional caucuses" (Africa; LAC; Asia-Pacific; Europe; North America; Western Asia and the Middle East; Arab Countries); "thematic caucuses and working groups" (Cities and local authorities; Community Media Caucus; Cultural and Linguistic Diversity; E-Government/E-Democracy; Education, Academia and Research; Education and Academia LAC Caucus; Environment and ICTs WG; Finance Caucus; Health and ICT Working Group; Human Rights Caucus; Indigenous Peoples Caucus; Implementation and Follow-up WG; Internet Governance Caucus; Media Caucus; NGO Gender Strategies WG; Patents, Copyright and Trademarks WG; Persons with disabilities; Privacy and Security WG; Scientific Information WG; Trade Union Caucus; Telecentres; Values and Ethics WG; WG on Volunteering and New ITs); and "multi-stakeholder caucuses (Gender Caucus; Youth Caucus). See Civil Society Meeting Point Website. http://www.wsis-cs.org/caucuses.html

struggled to gain recognition and influence. In some cases, the dynamism or lack thereof of various caucuses was affected by the ebb and flow of their key issues up and down the political agenda of the governmental negotiations. Another factor, however, was the gradual emergence of a community of practice within WSIS CS. This meant that the attention and involvement of many CS delegates tended to follow certain participants to the caucuses they coordinated or participated in and marginalize other caucuses whose leadership and members were not as well connected to the main CS networks. As the status, influence and membership of certain caucuses increased over the course of Phase II, seemingly at the expense of others, controversy emerged within the CS caucus system. This was particularly true with respect to the CS Internet Governance Caucus.

From Structures to Substance

In sum, the WSIS promoted a limited multi-stakeholder approach to policy-making at the supranational level. By allowing for the participation of various non-governmental stakeholders in summit deliberations, WSIS organizers sought to enrich the summit with the experiences, perspectives, and competencies of these stakeholders as well as reinforce the legitimacy of a political event that was largely otherwise disconnected from citizen control of decision-making. However, throughout the preparatory processes of both WSIS phases, this inclusive approach ran up against the objections of certain government delegations not interested in seeing multi-stakeholderism implemented concretely in the summit process.

Official texts and resolutions adopted at the WSIS became the main tools used by civil society actors to acquire extensive access and participation to key sites and political actors. These texts not only legitimized the participation of civil society organizations in the official political proceedings, but also transformed the political dynamic of the summit from intergovernmental to multi-stakeholder. As such, they became "ramparts" from which civil society organizations could confront the reluctance and resistance of certain government delegations towards the inclusion and participation of non-state actors in the WSIS. Once the WSIS multi-stakeholder principle became formalized in international resolutions, civil society was not willing to let it be abandoned by the summit.

• CHAPTER TWO •

Advancing Through the Phase II Preparatory Framework

The preparatory process leading up to the Tunis Summit began immediately following the conclusion of the Geneva phase. The process proceeded informally at first, in line with the Plan of Action adopted in Geneva.[1] It aimed at satisfying the demands of the United Nations General Assembly, which called for the event to be held in two separate phases and for official negotiations still undecided from the first phase to be settled in Phase II. An additional objective for Phase II was agreement on an implementation plan for the decisions taken at the WSIS. This pressure emerged, in response to newly passed resolutions in the UNGA that mandated a greater, clearer focus on following up and implementing the decisions taken by high level UN meetings, World Summits in particular.[2] The Tunis phase was to reflect the multi-stakeholder spirit and practice of the Geneva phase.

There was overlap, but the context had effectively changed from Phase I, and civil society had to reconfigure and adapt. Many sectors of civil society were dismayed and uncomfortable with Tunisia's role as host country for Phase II, justifiably as it turned out. In addition, having efficiently and successfully contributed to the official and informal proceedings of the first phase, civil society faced challenges related to maintaining a minimal level of meaningful participation in the political negotiations and sought in general to maximize the quality and scope of its participation. Finally, civil society in the second phase sought to ensure that the legacy of the WSIS would endure. If civil society could effectively influence decisions taken, a more solid foundation for a just, inclusive, and fair information society could be established. Civil society was also committed to ensuring that the implementation mechanisms, forums, and institutions arising from the WSIS would hold governments and the UN accountable to decisions taken at the WSIS and that concrete, on-the-ground measures would be put in

---

1   See paragraph 29 of the Geneva Plan of Action regarding the development of the second phase of the WSIS. http://www.itu.int/wsis/docs/geneva/official/poa.html
2   The UNGA resolution in particular, and its effect on the WSIS are discussed in more detail in Chapter 6.

place as a result. Finally, as they were conscious that the WSIS represented a new precedent in global governance, civil society actors strove to contribute to official proceedings in a useful and appropriate way, establish legitimate and effective internal organization mechanisms and ensure that the multi-stakeholder principle was included in all summit outcomes.

YEAR 2004

- 24-26 June—First Meeting of the Preparatory Committee (PrepCom-1), Hammamet (Tunisia)
- 20-21 September—Consultations on the establishment of the Working Group on Internet Governance (WGIG), Geneva (Switzerland)
- 4 October—First Meeting of the Task Force on Financial Mechanisms (TFFM), New York (USA)
- 22 October—First Meeting of the Group of Friends of the Chair (GFC), Geneva (Switzerland)
- 15-16 (afternoon) November—Second Meeting of the Group of Friends of the Chair (GFC), Geneva (Switzerland)
- 16 (morning) November—Consultation on Financial Mechanisms, Geneva (Switzerland)
- 16-18 November—Second Bishkek-Moscow Regional Conference on the Information Society, Bishkek (Kyrgyzstan)
- 22-23 November—Regional Conferences of Western Asia, Damascus (Syria)
- 23-25 November—First Meeting of the Working Group on Internet Governance (WGIG), Geneva (Switzerland)
- 29 November—Last Meeting of the Task Force on Financial Mechanisms (TFFM), New York (USA)
- 16-17 December—Third Meeting of the Group of Friends of the Chair (GFC), Geneva (Switzerland)[3]

YEAR 2005

- 10-11 January—Fourth Meeting of the Group of Friends of the Chair (GFC), Geneva (Switzerland)
- 2-4 February—Regional Conference of Africa, Accra (Ghana)
- 14-18 February—Second Meeting of the Working Group on Internet Governance (WGIG), Geneva (Switzerland)
- 17-25 February—Second Meeting of the Preparatory Committee (PrepCom-2), Geneva (Switzerland)
- 18-20 April—Third Meeting of the Working Group on Internet Governance (WGIG)
- 8-10 May—The Pan Arab Conference on WSIS–Phase II; An Arab Regional Dialogue, Cairo (Egypt)
- 31 May-2 June—Regional Conference of Asia-Pacific. Tehran, (The Islamic Republic of Iran)
- 8-10 June—Regional Conference of Latin America and the Caribbean, Rio de Janeiro (Brazil)
- 14-17 June—Fourth Meeting of the Working Group on Internet Governance (WGIG)
- 19-30 September—Third Meeting of the Preparatory Committee (PrepCom-3), Geneva (Switzerland)
- 13-15 November—Thirst Meeting of the Preparatory Committee (PrepCom-3a)
- 16-18 November—Second Phase of the World Summit on the Information Society (WSIS), Tunis (Tunisia)[4]

Table 1: Detailed Timetable of Official Events and Preparatory Process[5]

---

3 This is a modified version of the timeline provided at the WSIS official website. http://www.itu.int/wsis/preparatory2/calendar.html
4 Ibid.
5 Adapted from the WSIS official website, *WSIS Timetable for the Second Phase*. http://www.itu.int/wsis/preparatory2/calendar.html

This chapter will provide a step-by-step presentation of the issues, stages, and difficulties associated with each of these themes.

December 2003–June 2004: Reconstruction and Remobilization

The conclusion of the first phase of the WSIS in December 2003 marked the beginning of a period of uncertainty for civil society actors participating in the summit. In response to this uncertainty, various civil society actors mobilized around the issues and themes of the second phase and efforts were undertaken to convince summit organizers and national delegates of the necessity of fully including civil society in the organizations and activities of Phase II.

These demands were more or less achieved but were met with friction and push-back. As most of the issues addressed by the WSIS had been settled in the texts prepared in Geneva, a variety of WSIS stakeholders seemed to find it difficult to grasp precisely what the substance of the Phase II political negotiations was going to be.

The conclusion of the first phase was also followed by a politically and institutionally frantic period for the WSIS. The Civil Society Division was dismantled due to a lack of funds, and the Executive Secretariat of the WSIS had its management changed and personnel reduced.[6]

Adding to this uncertainty was the lack of clear organizational specifications for the preparatory phase of Phase II. The time, place, and manner in which the PrepComs were going to be held were relatively unclear at the start of Phase II. Thus, precise opportunities for civil society participation in the process were difficult to identify. Furthermore, some civil society organizations were wary of remobilizing to participate in the second phase, in particular, given that the Geneva phase had occupied three years of their attention and, in the eyes of many, had produced disappointing official results. The Tunis Summit was, virtually from the start, branded as a "summit of solutions," whose role was to transform the political decisions taken in the first phase into

---

6   Pierre Gagné, Executive Director of the WSIS Executive Secretariat during Phase I, was eventually replaced by Charles Geiger. Geiger first became Assistant Executive Director and worked under the direction of the Secretary-General of the ITU, Yoshio Utsumi, who initially assumed the managing role of the Summit. Geiger was confirmed as Executive Director in November 2004. Informal discussions revealed that the ITU was, generally, looking for ways to spend less money from its own budget on the second phase of the WSIS than it had on the Geneva phase. See, for example, Sally Burch, *Informal CS Bureau Meeting, ITU, 26 February 2004*. [WSIS CS-Plenary]. (February 26, 2004).

concrete initiatives to be implemented in practice. Yet, some of the more specialized civil society organizations that had taken part in the Geneva phase primarily because they perceived that its agenda touched on their specific issue areas were of the opinion that many such issues (communication rights, free software, media, etc.) had—in the eyes of the WSIS at least—either been resolved or effectively abandoned by the conclusion of Phase I. In other words, as the relatively open agenda of the WSIS had narrowed over the course of Phase I, many civil society actors' motivations to participate in the process dwindled.

Moreover, despite the fact that the Geneva Plan of Action clearly called for the establishment of multi-stakeholder working groups on information society financing and Internet governance, the candidate nomination processes, structures and participation methods of these groups were not in place by the start of the second phase. Nor were there clear general public indications of what role CS would play within them. This uncertainty added to the climate of cynicism developing in relation to the participation prospects for civil society organizations during the second phase.

In addition, the political climate in and reputation of Tunisia added to a feeling of uncertainty that prevailed at the launch of second phase operations. Civil society organizations feared police repression against any activists who were critical of the democratic and human rights record of the Tunisian regime.

At this stage, civil society's remobilization for Phase II focused primarily on two cross-cutting issues related to its participation and inclusion in WSIS proceedings.

The first aim was reaffirmation of the central character of the multi-stakeholder principle in the bodies, procedures, and documents of the Tunis Phase. For civil society organizations, this demand included: guarantees of full participation in preparatory committees and meaningful inclusion in the Working Group on Internet Governance and the Task Force on Financial Mechanisms and, reaffirmation of the importance of the multi-stakeholder approach in all Tunis documents.

In addition, CS sought to procure the maximum logistical and financial assistance available for the second phase. With Phase II being held in Tunis, it would be easier, in theory, for African civil society groups and organizations to participate extensively. But, the involvement of many CS organizations in the WSIS was contingent on the provision of some form of financial assistance. Civil society organizers suggested that the multi-stakeholder principle should be applied to the general financial aid made available to WSIS delegations, thereby ensuring that CS organizations from less developed regions would be represented. Based on the experience of Phase I, CS was clear that the provision of logistical resources—access to meeting space, translators, office equipment, administrative and technical support, etc.—was equally necessary in supporting the contribution of civil society to Phase II.

As the substantive political agenda for Phase II was not highly developed at this point, objectives, issues, and procedures of the second phase were still in the processes of being clarified. Thus, at this stage of the preparatory process, the objective of most civil society organizations was to ensure that the level of participation granted to them for Phase II did not fall below the levels that were achieved at Phase I. Civil society, that is to say, remobilized around the substantive issues and themes of Phase II much slower than it had in Phase I where various sectors of civil society—catalyzed early on by the CRIS campaign in particular—began working to push certain substantive issues onto and up the WSIS agenda virtually from the moment they got their feet in the door of the UN offices in Geneva.[7]

Organizing Phase II

The activities that led to the organization of the second phase actually began nearly three months after the conclusion of the Geneva Summit. The preparatory process initially began on March 3 and 4, 2004, with an informal stakeholder consultation meeting regarding the Tunis Summit.[8] Approximately one hundred participants from three different categories of actors at the WSIS participated in the event. Themes discussed at this meeting included: implementation of the Geneva Plan of Action, the process to be adopted for the second phase and its intended results and outcomes.

This meeting was significant because it effectively discussed and clarified a number of key points for the second phase and set the process down a certain path. However, in regard to CS, only members of the Civil Society Bureau were invited to attend. With the collapse of the CS Division, the Bureau was, by Phase II, the only civil society body to be officially recognized as a WSIS structure (see previous chapter). This episode re-ignited controversies surrounding the role and nature of the Bureau within civil society and renewed suspicions that it could be unrepresentative, inefficient, and relatively opaque in its operations. It also underlined that other,

---

7   See Marc Raboy and Normand Landry, *Civil Society, Communication and Global Governance: Issues from the World Summit on the Information Society*. New York: Peter Lang, 2005.
8   Tunisia and the ITU, *Éléments de rapport sur les activités de la réunion informelle.* (March 2-3, 2004). http://www.smsitunis2005.tn/plateforme/docs/reunioninformelle1.doc

more open and representative CS structures had to be formalized so that WSIS organizers and other stakeholders would understand and accept them as legitimate organs of CS participation.

Regardless, the discussions held at the March 2004 meeting clarified a number of key issues and took steps to determine some important guidelines for the process of Phase II, including responsibilities of the various stakeholders, how implementation of the Geneva outcomes would be accomplished, how Tunis phase documents would be drafted, and how follow-up on and evaluation of the results of the WSIS would be planned.

In parallel, the preparatory activities for the Working Group on Internet Governance also began during this time period (these are discussed in greater detail in Chapter 5).

The Road Towards PrepCom I

Details on the preparatory stages of the second phase of the WSIS were determined when the Intergovernmental Bureau convened at a closed-door meeting in March 2004. In addition to settling on its own composition for the second phase and confirming the availability of the preparatory committee president, this meeting set the time and place for the first preparatory committee meeting of Phase II.[9]

On March 25, 2004, Markus Kummer, a Swiss diplomat who had figured prominently in the Swiss government efforts to broker a last minute compromise before the Geneva Summit, was appointed head of the Executive Secretariat of the WGIG by the Secretary-General of the United Nations. The Secretariat was eventually established at the UN Palais des Nations in Geneva. The initial reaction of civil society members towards this appointment was very positive. Kummer had demonstrated a clear desire to operate along the principles of inclusivity and transparency, from the onset of the activities that led to the establishment of the WGIG Secretariat. In May 2004, Kummer announced that the working group would be fully operational by October. In the interim, the WGIG Secretariat would have to define procedures for selecting working group members, consider its composition and the roles and responsibilities of various stakeholders and define its operational processes.

---

9   Liliane Ursache, *Info on Phase II–Meeting of Intergovernmental Bureau*. [CS Bureau]. (April 8, 2004).

PrepCom I: Hammamet, 24–26 June 2004

The opening of the first PrepCom on June 24, 2004, in Hammamet, Tunisia, marked the end of the period of uncertainty that had characterized the organization of the second phase. It was suggested that:

1. The focus of the Tunis Phase should be:

    - Follow-up and implementation of the Geneva Declaration of Principles and Plan of Action by stakeholders at national, regional and international levels, with particular attention to the challenges facing the Least Developed Countries;

    - Consideration of the report of the Task Force on Financial Mechanisms (TFFM) and appropriate action;

    - Internet governance: consideration of the report of the Working Group on Internet Governance (WGIG) and appropriate action;

2. The agreements reached in the Geneva phase should not be reopened;

3. The output of the Tunis Phase should be a final document or documents, comprising a concise political part and an operational part, both of which reflect the areas of focus of the Tunis phase and reaffirm and enhance the commitments undertaken in the Geneva phase;

4. The preparatory process of the Tunis Phase should be inclusive, efficient, transparent and cost-effective; in principle, following the roadmap illustrated in the annexed chart (see following page).

Delegations from 127 governments and numerous non-governmental groups and individuals attended PrepCom I. As was the case with Phase I, the first preparatory committee meeting for Tunis planned the terms of the summit, its general objectives, and the deadlines for its work. The decisions taken in Hammamet set up, in other words, how the second phase of the WSIS would unfold. According to the schedule created in Hammamet, the TFFM report was due to be handed in at PrepCom II and the WGIG would have until the third PrepCom to submit its report to the summit. Thus the pressure was initially placed on organizing the TFFM and developing its work.

Figure 2: Outline of the Preparatory Process for the Tunis Phase of WSIS[10]

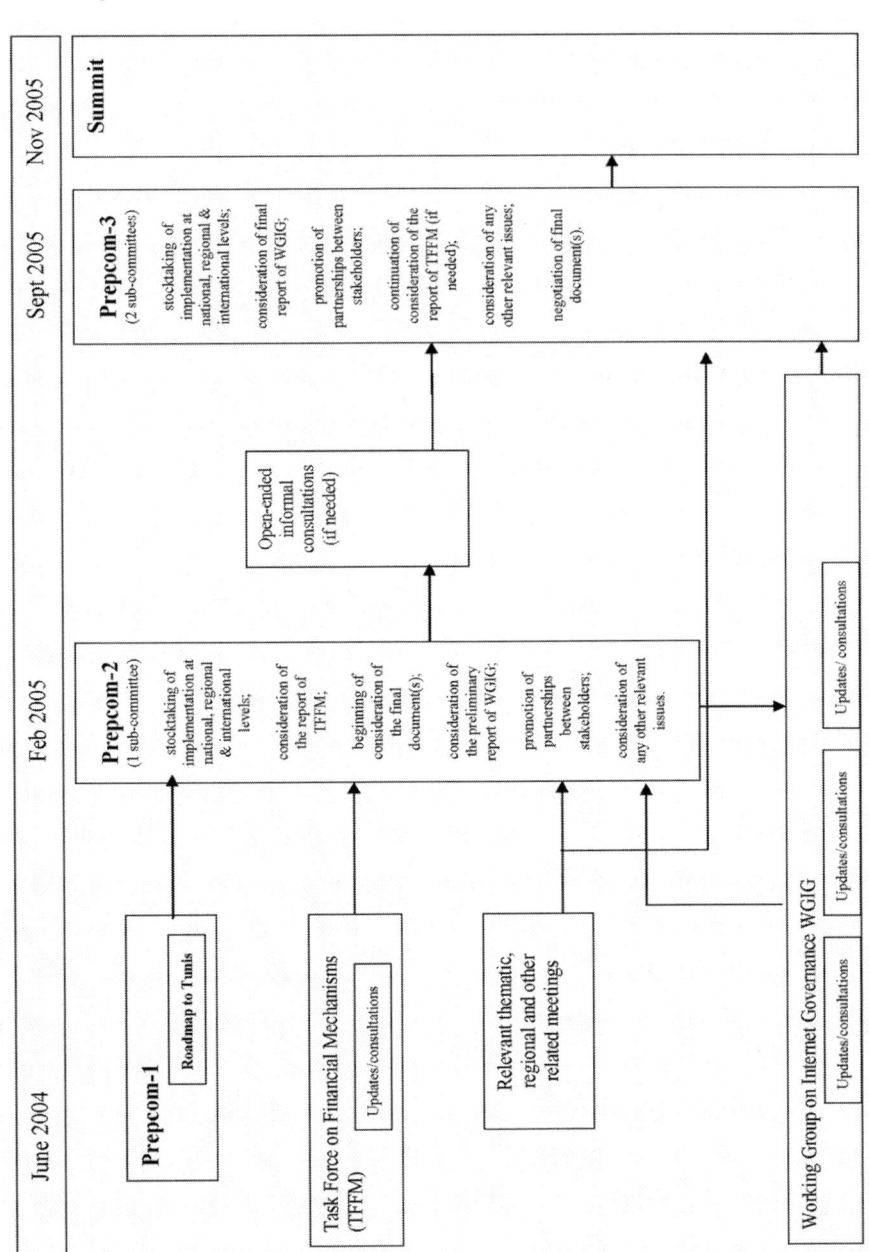

10 WSIS Executive Secretariat, *Decision of PrepCom-1* (WSIS-II/PC-1/DOC/5). (June 26, 2004). http://www.itu.int/wsis/docs2/pc1/doc5.doc

For CS in particular, a great deal hinged on the appointment of a new president of the preparatory committee, a decision to be publicly unveiled at PrepCom I. The replacement of the Phase I president of the preparatory committee, Malian diplomat Adama Samassékou, created uncertainty for CS. Samassékou was largely perceived to be a strong ally of civil society and a staunch supporter of the multi-stakeholder governance principle. Furthermore, his deft diplomatic touch and negotiation skills were seen to have influenced the degree to which certain government delegations were accepting of the active participation of non-governmental stakeholders during Phase I. There was no guarantee that the highly political process through which the new PrepCom president was chosen would lead to the nomination of a second individual in possession of similar traits. Furthermore, CS expected the PrepCom president to instill new life into the negotiations, help prepare discussions, and generally contribute to finding common ground between stakeholders and viewpoints. However, in the aftermath of PrepCom I, Latvian Ambassador Janis Karklins—who had been chosen to succeed Samassékou—immediately took steps to meet with civil society and to underline his commitment to its equal partnership in the second phase of the WSIS.[11]

Civil Society at PrepCom I

Civil society organizations used PrepCom I as a platform for defending the multi-stakeholder vision of the WSIS and to push for its application to all WSIS bodies. These concerns were summed up in an important intervention made before the Intergovernmental Plenary on June 25, 2004, by German privacy activist Ralf Bendrath on behalf of the Civil Society Plenary. Suggesting that governments had, over the course of the WSIS, already acknowledged that "governments can not address these issues alone" and that "any mechanism that does not closely associate civil society and other stakeholders is not only unacceptable in principle, it is also doomed to fail," Bendrath demanded that "the multi-stakeholder process be treated not just as "a nice phrase," but "becomes true reality." From there, a number of demands were made of the second phase of the WSIS, including:

- the ability to intervene in negotiations as points of interest come up, rather than at predetermined and often inappropriate times;
- clarification of how CS could contribute to the implementation of the action plan based on its unique knowledge and experience;

---

11 See Rik Panganiban, CONGO *Meeting with Karklins & Kummer Today.* [WSIS CS-Plenary]. (August 27, 2004).

- meaningful remote participation for global CS actors not in attendance;
- levels of inclusiveness and participation going beyond those that were granted during the first phase, or, at a very bare minimum, not rolled-back from the highest levels achieved during Phase I;
- mechanisms to ensure that civil society would be truly involved in any drafting processes and supported in commenting and proposing amendments in a timely manner during Phase II;
- modalities ensuring meaningful civil society consultation and cooperation on all areas of the stock-taking exercises and the implementation of the action plan;
- openness of all thematic and regional meetings to all interested stakeholders;
- transparent and equitable distribution of subsidies available to fund WSIS participation.

Going as far as threatening that CS's "further participation" in the WSIS would be dependent upon these conditions being met, Bendrath concluded with a sternly worded warning that "we are not willing to play an alibi role or lend our legitimacy to a process that excludes us from true meaningful participation. The summit can only be a summit of successes if there is substantive progress in our participation."[12]

Emphasizing issues in participation during the first PrepCom was a strategic move taken by civil society organizations in the aim of maintaining and leveraging the gains they had made over the first phase of the summit and consolidating their position at the WSIS. This emphasis on participation also pressured government delegations to implement the multi-stakeholder principle in the working groups that were being planned at the time and that, it was clear, would be crucial to framing the political negotiations of the second phase. Civil society fought especially hard to have the multi-stakeholder principle reflected in the WGIG.[13]

Yet despite these emboldened outward expressions of confidence in multi-stakeholder global governance, the location of the first PrepCom in Hammamet, Tunisia, put the internal participation structures of civil society under pressure. By the conclusion of the meeting, it had become more clear than ever that civil society was confronting serious internal issues. The relocation of

---

12 Ralf Bendrath, *Statement to the PrepCom Plenary on Behalf of the Civil Society Plenary*. (June 25, 2004). http://www.itu.int/wsis/docs2/pc1/plenary/heinrich-boll.doc

13 UN SC Working Group on Internet Governance, *Statement on the UN SC Working Group on Internet Governance Presented On Behalf of the Civil Society Internet Governance Caucus*. (June 26, 2004). http://www.itu.int/wsis/docs2/pc1/plenary/caucus-ig.pdf

the preparatory committee from Europe (where preparatory meetings had been held over the course of Phase I) to Tunisia created the opportunity for local CS actors to get involved and to integrate into existing CS structures. This was on the face of it a positive prospect, in particular for a summit that branded itself as being about development and bridging the digital divide. In practice, by far the largest contingent of new participants were drawn from groups with strong ties to Tunisia's authoritarian regime whose main aim appeared to be the disruption of the activities of other civil society organizations, in particular those working in the area of human rights. The overall objective of these groups, as it played out, was to prevent CS and the summit itself from adopting consensual texts critical of the Tunisian government's record on human rights. This had a serious destabilizing impact on those CS groups who had ambitions to use the WSIS to promote free speech and press freedom or at least to hold the Tunisian government to account.

A group of Tunisian participants presenting themselves as civil society delegates contested the designation of Souhayr Belhassen, from the Tunisian League for Human Rights, as the authorized civil society speaker to the Intergovernmental Plenary on June 25, 2004. The openness of civil society structures and the lack of explicit recognition of who can speak for it during meetings meant that these participants were able to contest the content of the speech that Belhassen was to read to government delegations by claiming that it did not represent their view and therefore should not be accepted as the input of civil society. This opposition, which was relayed by the Tunisian ambassador to the preparatory committee, blocked the plenary session for nearly an hour. One observer underscored the exceptional nature of this situation:

> One lateral but interesting aspect of the civil society debate around the human rights statement and who should read it was the very unusual circumstance of having a UN plenary meeting adjourn for around an hour in order to wait for Civil Society to put its act together! I can not recall anything like that in some 15 years of following UN meetings.[14]

An emergency civil society session was held to try to resolve the problem. This session too was destabilized by agitators who heckled participants and interrupted discussions. As a result of this intentional dysfunction, established CS processes could not operate and no decision was taken. The speech that was to be made before government delegates was cancelled, and it became evident that the strategy to infiltrate CS as a way of preventing criticism of the

---

14 Roberto Bissio (Quoted in Ralf Bendrath), *Fight About Civil Society Speaking Slots in Plenary: PrepCom Suspended Over Human Rights Dispute, Agents Provocateurs in Hammamet.* (June 25, 2004). http://www.worldSummit2003.de/en/web/638.htm

Tunisian government had worked and might continue to work. "If you look at the outcome," Ralf Bendrath wrote, "the Tunisian close-to-government 'NGOs' got what they wanted: There was no public civil society statement on human rights, and the woman from the Tunisian Human Rights League could not speak."[15]

The situation degenerated even further in the evening, and another CS delegate, Andy Carvin, described the subsequent drama this way:

> This evening's content and themes meeting of the civil society caucus degenerated into chaos, as some Tunisian and African NGO representatives overwhelmed the session, preventing chairs Karen Banks and Steve Buckley from leading a discussion on tomorrow's various civil society speeches to the government plenary. With probably seven or eight Tunisians for every non-Tunisian in the room, they demanded that civil society take an immediate vote on whether language critical of the Tunisian government would be excised from the human rights caucus text.
>
> The Tunisians, who did not participate in the human rights caucus session in which the language was drafted, demanded the right to overrule the text criticizing the Tunisian government, as well as change the speaker to someone they felt represented their view. They argued that a vote had been taken earlier in the afternoon during the previous civil society meeting—rather, it was their supporters shouting acclamation—and no consistent translation was offered to allow participants to make an informed decision.
>
> For nearly two hours, the audience of nearly 100 people were completely deadlocked, with the Tunisians blocking calls for a discussion proposing that two people one of their choosing and one chosen by the human rights caucus be given time to speak tomorrow during the government plenary. On numerous occasions, Karen Banks was shouted down by Tunisian representatives, saying she wasn't the legitimate chair of the meeting and that the chair that had presided over the chaotic afternoon session return to that position.[16]

Despite it being made clear that this situation threatened to derail CS participation in the PrepCom, the stalemate endured past eight p.m., when the translators signed off and any hope of facilitating an agreement was lost. The session thus ended with no resolution about subsequent steps for CS.

Souhayr Belhassen was ultimately invited to speak before the government plenary by the president of the PrepCom, in spite of the Tunisian objections. By way of compromise, speaking time was also allotted to the groups opposing her speech who, in a decidedly surreal scene, took advantage of this invitation

---

15   Ibid.
16   Andy Carvin, *Paralysis at Tonight's Civil Society Meeting.* (June 25, 2004). http://www.andycarvin.com/archives/2004/06/paralysis_at_to.html

to read a text that was identical to the speech given by Belhassen, except that it omitted all of the critical references she had made to Tunisia.

In the aftermath of PrepCom I, government delegations decided to hold the subsequent preparatory committee meetings scheduled for February and September of 2005 in Geneva rather than Tunisia.

This misadventure placed considerable pressure on civil society structures; the principles of openness, inclusion, and consensual participation could clearly be exploited and prone to capture by any large group with an agenda and a willingness to claim the civil society status that was, in principle, open to almost anyone. "In the light of these difficulties," it was determined, "civil society will be reviewing its practices and procedures in order to ensure that the diversity of opinions and perspectives that we represent can continue to be freely expressed."[17] Meanwhile, under the premise that "We have moved into the second phase of the WSIS, it seems necessary to consider the way forward," a sub-group of the Civil Society Bureau was formed at PrepCom I to review the composition and mechanisms of the CSB.[18] The Working Methods Working Group (WMWG), a second parallel, more involved and comprehensive program of evaluation and reform of CS structures, would be launched soon after (and will be discussed later in this chapter).

The first PrepCom thus tested the internal capacity of civil society organizations to manage internal disagreements that persisted over key summit issues. It also raised a fundamental question that extends beyond the context of the summit: who has the right to speak in the name of civil society?

A First Interlude: June 27, 2004–February 16, 2005

The period between the first and second PrepComs was crucial for the organization of the Phase II. The main issues addressed during that time were the following:

- composition and working methods of the GFC;
- establishment of the general principles for stocktaking;
- composition, operation, and launching activities of the WGIG;
- composition, launch, and completing the report of the TFFM.

---

17 Renata Bloem (quoted in Karen Banks and Steve Buckley), *Statement Prepared by the Coordinators of the WSIS Civil Society Content and Themes Group*. (June 26, 2004). http://rights.apc.org.au/news/2004/06/press_statement_cs_content_and_themes_coordinators.php

18 Viola Krebs, *Synthesis –CS Bureau Meeting–26/06/04*. [WSIS CS-Plenary]. (June 29, 2004).

Composition and Working Methods of the GFC

Government delegations adopted the following decision during the first meeting of the preparatory committee, held in Hammamet from the 24th to the 26th of June, 2004:

> The output of the Tunis Phase should be a final document or documents, comprising a concise political part and an operational part, both of which reflect the areas of focus of the Tunis phase and reaffirm and enhance the commitments undertaken in the Geneva phase; A group of friends of the President of the PrepCom of the Tunis Phase, with the assistance of the WSIS Executive Secretariat and in consultation with regional groups, will prepare a document to serve as a basis for negotiations in Prep-Com-2, taking into account, as appropriate, the outcomes of relevant thematic, regional and other WSIS-related meetings.[19]

The Group of Friends of the Chair was tasked with defining the format of the Tunis phase outcome texts and then proposing the documents and drafts to be used to facilitate negotiations. Given its agenda-setting potential, participation in the GFC was seen to hold decisive strategic importance to civil society's ambitions to meaningfully participate in Phase II. However, the GFC was established as a strictly intergovernmental process and CS overtures to these ends were flatly refused.[20]

The first GFC meeting took place on October 22, 2004, in Geneva. Open to observers from civil society and the private sector, the meeting planned the final composition of the group, its working methods, and the nature of its work. It also proposed a schedule for the group's activities until the second PrepCom, and declared that the GFC would operate out of Geneva.

The GFC was given a mandate of etching out a draft structure for the eventual final documents of the Tunis phase. The GFC would submit this framework to the preparatory committee, and the political negotiations of the entire preparatory process leading up to Tunis would be based around it. The general guidelines for the work of the GFC suggested that negotiations at Phase II of the WSIS should focus on development issues, complement the work done and agreements reached by the international community during Phase I, and avoid reopening any consensual agreement reached in Geneva. In terms of eventual outcomes of the Tunis phase, it was suggested that the GFC

---

19   WSIS Executive Secretariat, *Decision of PrepCom-1* (WSIS-II/PC-1/DOC/5). (June 26, 2004). http://www.itu.int/wsis/docs2/pc1/doc5.doc

20   For the structure and organization of the GFC and its membership see Janis Karklins, *Group of the Friends of the Chair: Outline of Composition of Group and Planned* activities. (August 31, 2004). http://www.worldSummit2003.de/download_en/Group-of-Friends-of-Chair-outline-31-August-2004.rtf. For the eventual composition of the GFC see the WSIS official website, *List of Members of the GFC* at http://www.itu.int/wsis/gfc/members.html

develop final documents that would reflect the structure of the Phase II Prep-Com process and adopt a two-part structure centered around a "political chapeau" section and an operational part. The political chapeau would be "a concise statement on the determination of Member States and other stakeholders in the WSIS process to develop and implement an effective and sustainable response to challenges and opportunities of building a truly global Information Society." Its aim was to:

- stress the growing importance of building an inclusive and development-oriented Information Society and outline new challenges deriving from this task,
- reaffirm the political will to move forward in bridging the digital divide, promoting digital opportunities, and
- reaffirm and enhance the commitments undertaken at the Geneva phase of the Summit as contained in the Declaration of Principles and the Plan of Action.[21]

Reflecting the focus of the preparatory process, the following structure was suggested for the operational part of the documents to be drafted by the GFC:

Chapter I: **From words to actions: a Summit of sustainable solutions**

- The work done by all stakeholders in implementing Geneva decisions
- Definition of appropriate mechanisms for the implementation of the Plan of Action

Chapter II: **Financial mechanisms for meeting the challenges of ICT for development**

- Possible improvements and innovations of financial mechanisms (reflecting the recommendations of the Task Force on Financial Mechanisms)

Chapter III: **Internet governance**

- Proposals for action (reflecting the recommendations of the Working Group on Internet governance)

Chapter IV: **The way ahead**
- Proposal on modalities of follow-up and review to the WSIS process[22]

The work of the GFC continued for nearly a year until September 2005. The group met four times before PrepCom II, held in February 2005.

---

21 Janis Karklins, *First Meeting of the GFC, Annex 3: Chair's non-paper to stimulate discussion in the GFC on October 22.* (October 22, 2004). http://www.itu.int/wsis/gfc/docs/1/annex3.doc
22 Ibid.

Stocktaking

On August 31, 2004, an informal meeting was organized in Geneva by the executive bodies of the summit in order to discuss the implementation and follow-up of the documents adopted during the first phase. A group of summit actors from different stakeholder groups was in attendance. The meeting also emphasized the need to establish measurable indicators and evaluate the contribution of the different parties working to reduce the problems identified at the WSIS.

On the basis of this meeting, the demonstrable and systematic progress that seemed to be possible through the straightforward effort to map the activities of and assign tasks and responsibilities to different WSIS stakeholders resounded with the Secretary-General of the ITU. A letter requesting all WSIS participants and stakeholders to fill out a follow-up questionnaire identifying and outlining engagements to date in WSIS-relevant issue areas and activities soon followed. The questionnaire was based on the 11 different themes defined in the Geneva Plan of Action.[23] We discuss these events and the stocktaking process that evolved from them in more detail in Chapter 6.

Task Force on Financial Mechanisms (TFFM)

The TFFM was officially launched in New York on October 4, 2004, with the goal of evaluating the capacity of existing international development financing mechanisms to effectively promote the use and incorporation of information and communication technologies. The TFFM was chaired by Mark Malloch Brown, an administrator at the United Nations Development Programme (UNDP).

Consultation was launched online in October 2004, and a consultation meeting open to all stakeholders took place on November 16, 2004. Though official rhetoric pointed to these meetings as evidence that the TFFM was guided by a spirit of openness and consultation, TFFM member appointment was top-down and opaque, the group worked under an extremely tight schedule and was given a very narrow mandate, and the report drafting work seemed to be undertaken even before the official launch of the TFFM process. A handful of prominent voices from WSIS CS were included in the TFFM process either as members, or in the various observer and expert intervener

---

23 See the Geneva Plan of Action. http://www.itu.int/wsis/documents/doc_multi.asp?lang=en&id=1161|1160

categories.²⁴ Yet, civil society would come to be highly critical of the limits that these constraints placed on its ability to meaningfully influence the TFFM procedures. These shortcomings were only accentuated by the inevitable comparisons that occurred between the TFFM and the parallel multi-stakeholder processes of the Working Group on Internet Governance.²⁵

The Launch of the WGIG: Composition and Working Methods

Consultations regarding the establishment of the WGIG took place in Geneva on September 20th and 21st, 2004.²⁶ More than 250 attendees discussed the composition, mandate, and working methods of the yet-to-be-established group that, according to the Geneva Plan of Action, would have to:

- develop a working definition of Internet governance;
- identify the public policy issues that are relevant to Internet governance;
- develop a common understanding of the respective roles and responsibilities of governments, existing intergovernmental and international organisations and other forums as well as the private sector and civil society from both developing and developed countries;
- prepare a report on the results of this activity to be presented for consideration and appropriate action for the second phase of WSIS in Tunis in 2005.²⁷

There was, according to Nitin Desai, Special Adviser of the Secretary-General on WSIS and the eventual chair of the WGIG, a "remarkable convergence of view on some key ideas." These included agreement that the WGIG should: approach IG from a broad perspective, taking into account what was already being done on IG elsewhere; be based on the multi-stakeholder principle; reflect a diverse array of backgrounds, regions, stakeholders, areas of specialty, and perspectives on the subject; and others.²⁸

---

24 For a complete list of TFFM members and other participants, see the final report (at page 95). Task Force on Financial Mechanisms for ICT for Development, *Financing ICTD–A Review of Trends and an Analysis of Gaps and Promising Practices*. (December 22, 2004). http://www.itu.int/wsis/tffm/final-report.pdf
25 The final report of the TFFM was published on December 22, 2004. The conclusions of the report and the issues associated with the process and civil society's critiques of it are reviewed in a following section of this chapter and further discussed in Chapter 4.
26 See the Working Group on Internet Governance website http://www.wgig.org/meeting-september.html
27 WSIS Executive Secretariat, *Geneva Plan of Action* (WSIS-II/PC-1/DOC/5-E). (December 12, 2003). Article 13 b (i, ii, iii, iv).http://www.itu.int/wsis/docs/geneva/official/poa.html
28 Distilled from Nitin Desai, *Consultations on the Establishment of the Working Group on Internet Governance, Chairman's Summary*. (September 20-21, 2004). www.wgig.org/docs/chairman-summary.pdf

The constitution of the WGIG became one of the most interesting procedural innovations of the WSIS. On November 11, 2004, the United Nations officially announced the WGIG's creation.[29] A level beyond the usual WSIS practice of subjugating non-governmental participants to the governmental delegations, the WGIG effectively established participation parity between government representatives, individuals based in the private sector, and representaves of civil society. All participated in the WGIG on equally footing in perhaps the only fully multi-stakeholder process convened during the WSIS.[30]

Open consultations on the WGIG's work took place on November 24, during the first WGIG meeting, which took place in Geneva from November 23rd to 25th, 2004. At this meeting, different organizational elements were discussed, including the working group agenda and the establishment of priorities and schedules. A list of themes linked to Internet governance was also compiled, and a series of working papers on these subjects was devised and assigned to various WGIG members.[31]

Regional Conferences

This period also saw the first three regional conferences of the second phase of the WSIS.[32] Each of these conferences produced a declaration outlining regional principles and priorities on issues related to the Phase II agenda to be incorporated into the WSIS as an official document. Civil society participation in these important meetings was virtually non-existent.

---

29 The process through which the CS contingent to the WGIG was selected and the controverisies that surrounded those decisions are discussed in Chapter 5.
30 United Nations, *Press Release PI/1620: UN Establishes Working Group on Internet Governance*. (November 11, 2004). http://www.un.org/News/Press/docs/2004/pi1620.doc.htm
31 See the WGIG website for *Inventory of Public Policy Issues and Priorities*: http://www.wgig.org/docs/inventory-issues.html. See also *Working Papers*: http://wgig.org/working-papers.html. The WGIG process is discussed in greater detail in Chapter 5 and, for a variety of in-depth accounts of the working methods of the WGIG written by WGIG members themselves see William Drake (ed.), *Reforming Internet Governance; Perspectives from the Working Group on Internet Governance*. New York: UN ICT Task Force, 2005.
32 The meetings in question were the following: Second Bishkek-Moscow Regional Conference on the Information Society: 16-18 November 2004, Bishkek (Kyrgyzstan), Western Asia: 22-23 November 2004, Damascus (Syria), and Africa: 2-4 February 2005, Accra (Ghana). For declarations/reports, see the WSIS official website: http://www.itu.int/wsis/documents/listing.asp?lang=en&c_event=s|2&c_type=co|ret

Civil Society's Reorganization and Evaluation of Its Participation in the WSIS

Motivated in part by the dysfunction that had been on display in Hammamet and growing frustration with the officiousness of the CSB, civil society reform activities began in the period between the first and second PrepComs. Various civil society organizations took advantage of a meeting of the UN ICT Task Force held in Berlin on November 19 and 20, 2004 to organize their own meeting at which plans for an evaluation of the organizational methods developed by civil society at the WSIS were hatched.

The Berlin meeting culminated in the development of one of the most promising initiatives devised by civil society during the second phase of the WSIS: the Working Methods Working Group (WMWG). The group's goal was to define a set of methods, procedures, and mechanisms that would, in conjunction with the Civil Society Bureau's own internal review process that had been launched at PrepCom I, strengthen the organization of civil society processes for the second phase of the WSIS and beyond.[33]

PrepCom II: 17–25 February 2005, Geneva (Switzerland)

Now back in Geneva, the second PrepCom distinguished itself from its predecessor in that it was the first negotiation meeting of Phase II that moved past questions linked to procedural issues and directly addressed the political themes of the summit as a primary topic of discussion. Civil society mirrored this dynamic by itself refocusing on issues of content and substance. In the process, PrepCom II underscored just how different CS participation in Phase II was going to be than it had been during Phase I. It became clear during PrepCom II that civil society participants held diverse and competing positions on substance, and, even in regard to many broadly defined issues, lacked the consensus and solidarity required to speak in a common voice to the extent that CS had done during the Geneva phase. As a result, efforts were made to draw to a greater degree on the caucuses and working groups, rather than from full CS structures such as C & T or Plenary, as sites where smaller-scale consensual positions could be developed and fed into the political negotiations.

Made acutely aware of instability involved in the plan to convene a multi-stakeholder summit in a territory controlled by an authoritarian regime by

---

33 The Berlin meeting, the WMWG and the general effort to reform CS process for Phase II is discussed in more depth in Chapter 3.

their PrepCom I experience, civil society began insisting at PrepCom II on discussion of the terms of the diplomatic immunity that would be granted to CS participants during the Tunis Summit through the WSIS host-country agreement.

Paradoxically, despite having experienced several major setbacks so far during the second phase—notably, marginalization within the TFFM and GFC—by PrepCom II, many activists were noticing an overall greater acceptance of the demands and propositions made by civil society. Though this degree of influence was highly contextual, civil society still managed to consolidate its position at the summit to a certain extent based on this newfound support. Government delegations, by now getting used to working with civil society at the WSIS, were progressively learning to appreciate the unique specialist knowledge and practical experience that CS had to offer on the issues being debated and to integrate the "expert" perspectives of CS into policy development and negotiation positions on WSIS issues.

Sally Burch, one of the most active civil society representatives at the WSIS, confirmed the change in attitude experienced during PrepCom II:

> The results of Phase I are clearly visible in terms of the greater openness of the official intergovernmental process to receive and consider civil society input. Many government delegations have actually been requesting civil society contributions to improve quality of the documents, the first drafts of which are extremely vague and general. In these circumstances, it makes sense to give priority to developing the input and getting it to governments in time. Broader consensus is likely to be needed further down the line as it becomes evident what the critical issues and areas of blockage are.[34]

This process proved to be somewhat cyclical, however. Expertise was in high demand at the beginning of the negotiation process. But the window of receptiveness to civil society input seemed to close gradually as negotiations progressed and government delegates shifted from a fact-finding and position-developing brainstorming phase into a negotiation phase. By later in Phase II, politics and time constraints had effectively limited the options that were available to government delegations and, in the process, diminished their need for, capacity to act on, and time to engage with, perspectives from CS.

The efforts of the GFC during the second PrepCom gradually converged into two working documents that were used as a basis for negotiation. These documents, entitled *Political Chapeau / Tunis Commitment as of 11 January 2005* and *Operational Part of the Final Document / Tunis Agenda for Action / Tunis Plan*

---

34 Sally Burch, *Report 1 from Geneva WSIS 2 PrepCom: Civil Society Reorganizing Around Content.* (February 22, 2005). http://www.movimientos.org/foro_comunicacion/show_text.php3?key=4084

of Implementation as of January 2005,[35] were prepared after consulting with actors that were invited to participate in the GFC. These documents were based on the GFC's efforts to synthesize and organize the various contributions made to the GFC process with the TFFM report, which was to form the basis of the sections regarding financing. These working drafts also reflected the WGIG's preliminary report which had been submitted on February 21, 2005. The negotiation sessions held at PrepCom II mostly concerned issues relating to the text on ICT financing for development and, to a lesser degree, Internet governance, the Political Chapeau, and the Implementation Plan of the Geneva Action Plan.

Discussions on Financing and the DSF

The TFFM report submitted on December 22, 2004, served as a catalyst for debates on the financing issue. In general, the report was received coldly by civil society organizations involved in the WSIS as well as by a group of developing countries. Many civil society organizations, together with several government delegations from the Global South, denounced the report's insistence on the neo-liberal principle of economic liberalization. Members of the TFFM themselves further criticized the emphasis placed on the private sector as a vehicle for development. Debates over the costs involved in reducing the digital divide and allocating financial responsibility for doing so further polarized negotiations. Developed countries, headed by the United States and the European Union, clearly voiced their desire to hold back from making new concrete financial commitments. This hesitation was met with great displeasure from the developing countries that had been hoping for the establishment of some form of binding financial-transfer mechanism that would force countries with high degrees of connectivity to increase their contributions to supporting connectivity in the developing world. In the end, there was a degree of rapprochement at the level of principles, as the respective roles of the private and public sectors in the integration of marginalized populations into global information and communication networks were at least acknowledged as was the voluntary Digital Solidarity Fund, which had been recently

---

35   See WSIS Executive Secretariat, *Draft Political Chapeau / Tunis Commitment as of 11 January 2005*. (January 11, 2005). http://www.itu.int/wsis/gfc/docs/4/political-chapeau11 jan-pm.html. See also WSIS Executive Secretariat, *Draft Operational Part of the final document / Tunis Agenda for Action / Tunis Plan of implementation as of 11 January 2005*. (January 11, 2005). http://www.itu.int/wsis/gfc/docs/4/operational-part11jan-pm.html

established in Geneva by local and municipal authorities.³⁶ By qualifying acknowledgment of some sense of responsibility to deal with the digital divide and the proposal of a mechanism with the caveat that there was no coercive measure in place that would obligate developed governments to contribute financially to the solution in any way, winks were made in the direction of both sides of the debate. Yet this was a fraught compromise and an ultimately empty gesture.

Internet Governance

The WGIG held a second meeting in parallel to the opening of the second PrepCom, from February 15 to 18, 2005. The second WGIG meeting featured debates on possible definitions for Internet governance. A preliminary report³⁷ was presented to PrepCom II and the debate that followed allowed for some sense of how the group's work might eventually be received at the WSIS. From there, the WGIG turned to evaluation of existing Internet governance mechanisms and to developing a "common understanding of the respective roles and responsibilities" of various stakeholders involved in these mechanisms.

By virtue of the co-equal treatment being given to CS members of the WGIG and the ability of those individuals to both draw from and report back to CS structures, during the WGIG period CS was as close to the action as it had been at any previous point during the WSIS. There was considerable enthusiasm about the influence that CS positions might be able to excercise on the subsequent negotiations and PrepCom II was the site of efforts to develop a base consensus on the CS position on IG. One outcome of these efforts was the drafting of a statement that pointed to the central and determining role of human rights in the governance process while insisting that IG must: be open, inclusive and transparent, and allow for equal participation of all actors in its institutions and processes; give consideration to issues linked to different types of economies and their development; and preserve and promote cultural and linguistic diversity on the Internet.³⁸ Another outcome of CS efforts around

---

36 See WSIS Executive Secretariat, *Final Report of the Preparatory Meeting (PrepCom-2 of the Tunis phase)* (WSIS-II/PC-2/DOC/12-E). (February 25, 2005). http://www.itu.int/wsis/docs2/pc2/off12.doc. See also further on in this chapter for details on the launch of the DSF and Chapter 4 for more in-depth discussion of the TFFM report and the DSF.

37 See WSIS Executive Secretariat, *Preliminary Report of the Working Group on Internet Governance (WGIG)* (WSIS-II/PC-2/DOC/05). (February 21, 2005). http://www.itu.int/wsis/documents/doc_multi.asp?lang=en&id=1460|0

38 See especially *Statement by the Civil Society Internet Governance Caucus, the Gender, Human Rights, Privacy and Media Caucuses on Behalf of the Civil Society Content and Themes Group.* (February 23, 2005). http://www.wgig.org/docs/CS-Hofmann.rtf. See also, Civil Society

Internet governance at PrepCom II, however, was the emergence of divisive conflicts between various sectors of CS that centered on accusations of agenda setting. As a result, beyond the sort of very broad parameters discussed above, in the aftermath of PrepCom II, the very notion that CS could come to a consensus view on issues related to Internet governance was seen to be increasingly fraught.[39]

Implementation, Follow-up, and the Organization of the WSIS

Of the three main issues that were raised during the second phase, those surrounding implementation and follow-up certainly received the least attention until later in the preparatory process. This is partially explained by the cart-before-the-horse problem: obviously it is difficult to map out follow-up and implementation processes before the major decisions on content are themselves made. But, its status as a low priority item also probably reflected a lack of enthusiasm on the part of many governments—in particular on the part of many OECD country governments—for making potentially expensive and binding concrete commitments to further action.

During the second PrepCom, implementation procedures remained vague and controversial. In reaction to this, the GFC proposed a general implementation structure, a form of which would eventually be adopted by the WSIS. According to this structure, every "action line" that was defined in the Geneva Plan of Action and the Tunis Agenda for the Information Society would be managed by a UN body (department, agency, etc.) operating under the multi-stakeholder principle of the WSIS. Each UN body would periodically prepare progress reports on the implementation process, which would then be sent to a coordinating body in charge of managing implementation and follow-up measures. This coordinating body would periodically compile all information obtained on these measures into a single report that would be submitted to the United Nations General Assembly.[40]

This approach was generally supported by civil society, which nonetheless called for more transparent, inclusive, and open implementation and follow-up mechanisms to be developed bottom-up rather than imposed top-down by

---

    Human Rights Caucus, *Human Rights and Internet Governance.* (February 23, 2005). http://www.itu.int/wsis/docs2/pc2/plenary/HUMANRIGHTS.doc

39 This episode, and the general issues related to CS involvement in the IG debate, are discussed in greater detail in Chapter 7.

40 Group of Friends of the Chair, *Draft Operational Part of the final document / Tunis Agenda for Action / Tunis Plan of implementation as of 11 January 2005.* (January 11, 2005). http://www.itu.int/wsis/gfc/docs/4/operational-part11jan-pm.doc

UN agencies. Such a demand was seen to be closer to the summit's multi-stakeholder principle than the alternative proposed.

Civil Society Views on Official Negotiations and Its Involvement in the WSIS Process

Second PrepCom negotiations on content got off on the wrong foot for many civil society participants. CS criticized the lack of vision and commitment on the part of political leaders, the political stalemate on the key issues, the strong ideological biases embedded in many proposed solutions, and the absence of adequate discussion on the implementation and follow-up mechanisms that were supposed to be planned during the second phase of the WSIS.

Following up on the demands voiced by Ralf Bendrath on behalf of the CS Plenary at PrepCom I, some CS participants began to question their continued involvement in the WSIS at PrepCom II. A series of strategies were on the table including an organized pull-out of some or all segments of WSIS CS from the official process, moving to resurrect, revise, update and expand upon the civil society declaration of Phase I, or even take steps to start drafting a second and entirely new parallel civil society declaration. Against the backdrop of this strategic discussion on the influence and relevance of continued civil society participation in the summit, the revision of CS's participation structures continued undaunted. By PrepCom II, the Working Methods Working Group had formulated a series of proposed changes that sought to ensure the neutrality of the CS plenary chair, create a charter defining decision-making processes for content and themes, create a mediation process in cases of disputes between various sectors of CS and create a more open, transparent and formal process for defining caucuses, working groups and their membership.[41]

A Second Interlude: February 26, 2005–September 18, 2005

During the period between PrepCom II and PrepCom III a number of parallel tracks began to merge:

- the Digital Solidarity Fund launched in March 2005 in Geneva;
- regional meetings convened for Asia-Pacific, Latin America, and the Caribbean in May and June;

---

41 Rik Panganiban, *Information Note on Working Methods Working Group at PrepCom II*. [WSIS CS-Plenary]. (February 24, 2005).

- the GFC held its final round of meetings and indicated the completion of its work by distributing a revised set documents for use as the basis of negotiations at PrepCom III;
- for the first time, meaningful debates occurred on implementation and follow-up measures, in relation to various drafts of the texts being prepared by the GFC;
- the third and fourth WGIG meetings were held in April and June 2005, and the final WGIG report was published in July 2005.

The Launch of the Digital Solidarity Fund

The Digital Solidarity Fund that was originally proposed during the first phase of the WSIS by Senegal as a mechanism for transferring money for ICT infrastructure funding from developed to developing countries was finally inaugurated on March 14, 2005. By then, what was originally devised as a tax on information and communication technology expenses that could help fund connectivity in the developing world had been downgraded to a voluntary fund that gained the backing of a group of cities and local authorities but had little support amongst OECD governments.[42]

Regional Conferences

The Asia-Pacific conference was held between May 31 and June 2, 2005, in Teheran, Iran. Civil society access to this meeting was restricted by organizers and overall CS participation was low. The individuals present expressed their dissatisfaction with both the way in which the meeting was conducted and the texts that were adopted.[43]

The Latin American and Caribbean regional conference was held between the 8th and 10th of June, 2005, in Rio de Janeiro, Brazil. This meeting was noticeably more open to the participation of civil society. Unlike the meeting in Iran, civil society organizations were actually invited to participate in this conference. The contrast between the inclusive stance towards civil society organizations at the Rio conference and the restrictive attitude displayed at the Teheran conference illustrates the extent to which regional differences can

---

42 The role and place of the DSF among the many development projects related to information and communication technologies will be discussed and analysed in greater detail in Chapter 4.

43 See especially Natasha Primo, *Iranian Authorities Ban West Asia and Middle East WSIS Civil Society Meeting on Kish Island (Iran)–Statement and Call to Action.* [WSIS CS-Plenary]. (September 1, 2005).

influence policies governing the participation of non-state actors in governmental working meetings. Yet, the official documents drafted in Teheran and Rio would go on to receive equal weight in subsequent negotiations.[44]

The WGIG Completes Its Work and Publishes Its Final Report

At its third meeting, the WGIG created more detailed definitions of the roles of Internet governance stakeholders and evaluated the efficiency and performance of existing Internet governance mechanisms. The WGIG considered options for further aligning such mechanisms with the Geneva Principles (i.e., democratic, transparent, multilateral, and fully open towards the participation of the private sector, civil society, international organizations, and governments, while at the same time remaining efficiently coordinated).

According to Wolfgang Kleinwächter, a CS-based member of the working group, it was at this point that broad acceptance was evident amongst WGIG that the difficult issues of Internet governance were not only those based on network resource management, but also cross-cutting issues such as spam, copyright, Internet security and others.[45]

The final WGIG meeting was held June 14–17, 2005, and the WGIG distributed a report on the consultation procedures, contributions, and opinions expressed on its work in June. The WGIG's final report (discussed in greater detail in Chapter 5) was distributed on July 14, 2005.[46] It proposed a working definition and four models of Internet governance. In the WGIG's working definition, Internet governance is

> the development and application by Governments, the private sector and civil society, in their respective roles, of shared principles, norms, rules, decision-making procedures, and programmes that shape the evolution and use of the Internet. (WGIG Report, page 4)

The WGIG's four proposed models of the institutional structure of IG were described in the following terms:

- One model sees no need for a specific oversight organization, but envisages the possibility of enhancing the role of the Governmental Advisory Committee (GAC) of the Internet Corporation for Assigned Names and Numbers (ICANN).

---

44 See the WSIS official website, *Outcome of WSIS Regional and Thematic Meetings.* http://www.itu.int/wsis/documents/listing.asp?lang=en&c_event=s|2&c_type=co|ret

45 Wolfgang Kleinwächter, *WGIG.* [WSIS CS-Plenary]. (April 21, 2005).

46 The WGIG's reports, along with any documentation compiled during its consultations, are available on the WGIG website. http://www.wgig.org

- Another model suggests setting up a new body that would address public policy issues in relation to ICANN competencies and maybe also issues that do not fall within the scope of other existing institutions. In this model, the GAC might be made redundant.
- A third model envisages the creation of a new body that would replace the GAC and have wide ranging policy competencies. ICANN would be accountable to this new body, which would also facilitate negotiation of Internet-related treaties, conventions and agreements. It would be linked to the United Nations.
- A fourth model proposes new structures for three interrelated areas of Internet policy governance, oversight and global coordination. It suggests the creation of three new bodies for each of these functions and would include a reformed internationalized ICANN linked to the United Nations.[47]

The WGIG also proposed the creation of a new organizational space for discussion on global Internet governance that would be open to all stakeholders. Presented as an Internet governance "forum," this space would allow for multi-stakeholder participation and discussion on issues, positions, and points of view concerning public policy issues linked to Internet governance.[48]

Civil society was generally very satisfied with the working methods adopted by the WGIG. For the most part, CS considered the WGIG's working procedures to be inclusive, participative, and transparent. The exception was some criticism that certain segments of CS had not been able to effectively communicate with WGIG members. The reactions of civil society organizations toward the content of the report were more varied, as no veritable consensus on Internet governance existed among them beforehand. That said, the proposal to create the IGF as a follow-up multi-stakeholder global governance organization represented a provocative possibility for the continued and semi-permanent role of CS in global governance of communication and a potentially precedent-setting move whose realization would ensure that CS's foot was further wedged into the door of the intergovernmental system. As a result, the IGF became a logical rallying point for CS support and ambitions to enhance the degree of multi-stakeholder participation in UN processes, even for CS WSIS participants not primarily or even actively engaged in the Internet governance debate.

---

47 Secretariat of the Working Group on Internet Governance, *UN Press Release: Independent Group Submits Report on Internet Governance in Lead-up to Tunis Summit on the Information Society*. (July 14, 2005). http://www.wgig.org/docs/PRESS-RELEASE-14.07.05.pdf

48 The reception of and politics around the IGF proposal at WSIS are discussed in greater detail in Chapter 5. The IGF is the focus of significant discussion and analysis in Chapter 7.

The GFC Completes Its Work:
Questions of Implementation and Follow-up

During PrepCom II, the GFC was tasked with the preparation of new proposals for the operational document being drafted for the Tunis phase. The group met in three closed meetings and one open meeting between June 27th and September 7th, 2005. An informal meeting for the GFC was also organized on June 13th in Geneva.

The June 13th informal meeting reaffirmed the WSIS's international development orientation and the importance of the multi-stakeholder principle for the summit's second phase as well as its implementation and follow-up mechanisms. This meeting also reaffirmed that the WSIS was not going to create a new UN agency strictly for the purpose of implementing its decisions and thus, that subsequent negotiations on implementation and follow-up should focus on mobilizing and otherwise making use of existing institutions.

Despite the preparatory committee president's desire to open-up GFC meetings to civil society, the Intergovernmental Bureau did not invite any civil society organizations to this meeting. Nevertheless, different civil society actors showed up to the meeting and did their best to gain some form of participation.

During its eighth and final meeting, the direction opted for by the GFC evidenced significant influence from proposals that had been made by PrepCom President Karklins. The result was a drastic change to WSIS plans for implementation and follow-up mechanisms.[49] Though the work of the GFC concluded here, this new direction would go on to become the subject of intense negotiations several weeks later during the third PrepCom.

PrepCom III: September 19-30, 2005, Geneva (Switzerland), and November 13-15, 2005, Tunis (Tunisia)

The third and final PrepCom of Phase II was marked by three important and interconnected events. These were:

- the end of government negotiations, which put the credibility of the WSIS to the test;
- the completion of civil society's reorganization efforts, followed by its preparation for the Tunis Summit;

---

49 See Chapter 6 for more details.

- amplification of the questions CS was asking itself about civil society's role in the work of the summit and within the multi-stakeholder global governance model.

PrepCom III: Geneva

The final PrepCom of both the Tunis Phase and the WSIS itself began on September 19, 2005. It did so amid palpable tension around the fact that crucial and highly controversial points at the centre of the government negotiations were emerging as topics of meaningful negotiation, just as the summit process entered its final preparatory meeting. Issues of Internet governance as well as implementation and follow-up mechanisms for WSIS decisions, in particular, were still very much open for debate.

Many civil society organizations believed that implementation and follow-up mechanisms would determine whether or not the summit could be labeled a success. For CS, the efficiency and impact of these mechanisms would reveal just how far governments were willing to go to solve the social and economic issues related to the emergence of the information society. Civil society wanted to see the WSIS as more than just a linear process leading to a declaration of intent. They envisioned it as a step towards ultimately reducing the inequality that had excluded entire segments of population from the most up-to-date communication networks through the implementation of concrete initiatives and the establishment of reliable indicators. Furthermore, civil society recognized the strategic importance of such mechanisms to its own participation in post-WSIS events. The many civil society organizations that had participated in formal WSIS proceedings for more than four years now refused to be excluded from the implementation and follow-up mechanisms that were going to be set up by the summit. This refusal to be excluded was reiterated many times and stemmed from the desire to keep the implementation and follow-up process linked to the multi-stakeholder principle and to ensure, going forward, that the global governance of the information society would not roll back the progress toward CS inclusion that had been made during WSIS.

The work at PrepCom III was divided between two subcommittees. Subcommittee A would negotiate an agreement between government delegations on Internet governance, and Subcommittee B would negotiate agreements between government delegations on everything else that remained to be resolved, primarily implementation and follow-up mechanisms.

Each subcommittee convened in 14 full meetings. In addition, the work given to Subcommittee A was further subdivided into drafting groups.

Internet Governance: Who Controls the Internet?

The negotiations on Internet governance truly began at the third PrepCom. From a political standpoint, the major Internet governance issue that remained unresolved was the control of root zone files. The ICANN, a nongovernmental institution responsible for attributing and coordinating domain names and IP addresses at the highest level, had been plunged into the middle of a charged debate. OECD governments, the private sector and the Internet technical community continued at PrepCom III to defend the status quo while Brazil, Saudia Arabia, Iran and other emerging and developing governments pushed for the WSIS to impose some form of intergovernmental institutional governance.

The third preparatory committee meeting opened with seemingly irreconcilable positions on the issue and would wrap up its planned meeting sessions with no consensus.[50]

Everything Else: Political Chapeau, Implementation, and Follow-up at PrepCom III

The political chapeau—which would eventually become the Tunis Commitment—remained highly controversial throughout the third PrepCom. Paragraphs touching on issues related to national sovereignty, human rights, the respective roles of stakeholders, trade and liberalization, debt relief for developing countries, the digital divide, and the mobilization of resources continued to be negotiated until the PrepCom resumed in Tunis on the eve of the summit in November 2005.[51] On follow-up and implementation, governments were unable to agree on the leadership, form, or structure that implementation and follow-up activities would assume after the WSIS.[52] Civil society was strongly critical of this weakness and began to consider government commitments on follow-up and implementation as something of a litmus test of the entire summit's credibility.

Civil Society at PrepCom III: Part One (Geneva)

The restructuring process for civil society structures, which had begun nearly a year earlier, came to an end with the closing of the third PrepCom. Reform

---

50 For a detailed account of the debate over IG at WSIS and the compromise that concluded it, see Chapter 5. For a reflection on the role of CS in the global IG debate at WSIS and since, see Chapter 7.
51 Conference of Non-Governmental Organizations in Consultative Relationship with the United Nations (CONGO), *Civil Society Orientation Kit*. (November 2005). http://www.Ngocongo.org/congo/files/wsis_oriention_kit.pdf
52 See Chapter 6 for details.

plans proposed by the Working Methods Working Group for the Civil Society Plenary and Civil Society Bureau were adopted by the CS Plenary on September 28, 2005.[53] The documents represented a positive initiative for the continued participation of non-state actors in the global multi-stakeholder governance process that could serve as a point of reference for subsequent meetings and procedures. Civil society also debated and established participation procedures for the Tunis Summit, including procedures for appointing speakers, granting any overpasses that would permit access to restricted-entry events and planning parallel activities.

Civil society's organization at the WSIS was closely linked to the conditions for participation that had been laid out for it in the different bodies, proceedings, and structures of the summit. The main plenary of PrepCom III was broken out into two governmental subcommittees. This represented a different structure than had been used during previous PrepComs. For the first part of PrepCom III, there was uncertainty from summit organizers about how non-governmental participation was going to work in subcommittee meetings, and confusion within CS about how contributions should be submitted and interventions made by various caucuses and working groups. It quickly became evident, however, that the work of the subcommittees had to be closely followed in order for civil society contributions to be effective because civil society organizations had to be able to respond promptly to the evolving language of the Subcommittees, as well as propose content that was in line with their most recent developments.

Civil society therefore decided to adopt working structures that would reflect those of the two subcommittees. The Internet Governance Caucus became the effective focal point of CS input to subcommittee A: following and reflecting the negotiations, drafting statements and nominating speakers to make interventions, sharing details with CS through email lists, CS Plenary and C & T meetings, and doing outreach to, and integrating comments and interventions from, other CS caucuses and working groups. A newly created working group on follow-up and implementation took on a similar focal point role for Subcommittee B.[54]

The negotiations of implementation and follow-up, and the roles granted to CS within the mechanisms devised therein, were framed as a barometer for

---

53   These reforms are detailed in the Civil Society Orientation Kit developed for PrepCom III and the Tunis Summit and are discussed in greater detail in Chapter 3 of this book.
54   See Robert Guerra, *Notes on Civil Society Bureau Meeting*. (September 18, 2005). http://wsis.civiblog.org/blog/WSIS/CivilSocietyBureau/_archives/2005/9/20/1242642.html See also Bertrand de La Chapelle, *Invitation to Join the Working Group on Sub-Committee B*. [Follow-up]. (September 20, 2005).

measuring the ultimate success or failure of the WSIS process. "It will define the real credibility of the entire work done by the Governments over the last five years," one CS intervention to Subcommittee B suggested, continuing that "The outcomes from the Tunis Summit on this issue will be the benchmark upon which the *real political will of the Governments* to implement decisions and to bridge the digital divide is measured" (emphasis in original). [55]

Civil society also devoted a large portion of its efforts and energy towards ensuring and expanding its own participation in evolving processes of Prep-Com III. The preparatory committee rules and procedures established during the first phase granted observers from civil society and the private sector access to any subcommittee established by a preparatory committee.[56] Civil society organizations were also given 15 minutes of speaking time in subcommittee meetings. However, the rules and procedures did not mention whether these actors could take part in drafting groups operating under the subcommittees, which created often frantic confusion when drafting groups were indeed eventually formed at PrepCom III.

Exclusion from these drafting groups—small breakout meetings of government representatives delegated with the responsibility to craft the exact wording used to articulate issues of consensus in official summit outcome documents—would directly undermine the ability of civil society organizations to not only have input on, but hold government delegations accountable for this crucial phase of negotiation. In the initial period of uncertainty created by the convening of the drafting groups, decisions about the inclusion of CS were often made by the appointed chair of each drafting group and reflected the feelings of the sample of individual government delegates that happened to be in the room at the time. Some government delegations took advantage of this ambiguity to exclude members of civil society from drafting committees entirely. Others opted for a "talk and walk" approach, where civil society organizations were told to leave the room after delivering their comments. Unofficially, however, certain drafting groups allowed civil society participants to be silently present at their meetings. As is often the case at high-level political events, the unofficial procedures were just as influential on the dynamics of the work and negotiations as the official ones. In this way, civil society was able to respond quickly to its formal exclusion from the drafting committees. But the initial lack of a clear official directive that CS should be included in

---

55 Civil Society Working Group on Follow-up, *Statement*. (n.a.). http://www.choike.org/nuevo_eng/informes/3656.html
56 WSIS Executive Secretariat, *Report of the First Meeting of the Preparatory Committee* (WSIS03/PREP-1/11(Rev.1)-E). (July 12, 2002). http://www.itu.int/dms_pub/itu-s/md/02/wsispc1/doc/S02-WSISPC1-DOC-0011!R1!MSW-E.doc

drafting groups unconditionally mobilized CS around calls for full partnership rights within the multi-stakeholder model.[57]

The third PrepCom was also marked by political tensions resulting from the refusal of summit organizers and government delegations to accredit Human Rights In China, a non-governmental organization based in the United States, for participation in the meeting. When the delegation from the United States asked the Executive Director of the WSIS secretariat, Charles Geiger, to explain this refusal, he cited issues related to the transparency of the organization's financing as the rationale. As China had become a strong opponent of acknowledging the group in all international events, several observers assumed that the Chinese delegation had been pressuring summit organizers to ensure that the group remained excluded from the WSIS. The United States demanded that the NGO be accredited. Discussions on this matter required the temporary suspension of the PrepCom at one point. Ultimately, government delegates were allowed to vote on whether or not the HRC should be recognized. The vote resulted in 55 votes for not accrediting the group and only 35 for accrediting it, with the rest of the delegations abstaining from the vote.[58]

PrepCom III Resumed: The Final Countdown, November 13–15, 2005 (Tunis)

During international political negotiations, political urgency often forces negotiators to come to a consensus. When faced with the possibility of political failure, government delegations are often obligated (not to mention pressured) to be conciliatory, as the desire to fully realize their wishes cedes to the need to adopt a position that at least partially reflects their interests in the face of a possible all-out political failure. In the case of the WSIS, political urgency manifested itself in veritable negotiation marathons that were held at an almost frantic pace. The urgency, however, had a positive outcome: consensus was reached on the issues that had not been resolved up until then. These issues were:

---

57 Discussed and reflected on in greater detail in Chapter 3.
58 The politics surrounding this episode underline an important element of the Cardoso report, which contains a detailed discussion of the need to establish uniform and transparent rules for the accreditation of CS organizations to UN processes. See Fernando Henrique Cardoso et al., *We the Peoples: Civil Society, the United Nations and Global Governance. Report of the Panel of Eminent Persons on United Nations–Civil Society Relations* (A/58/817). (June 11, 2004). p. 13. http://www.un.org/french/ga/search/view_doc.asp?symbol=A/58/817&referer=http://www.un.org/french/reform/panel.html&Lang=E

- the establishment of a Forum on Internet governance;
- "enhanced cooperation" on existing practices of Internet governance;
- the establishment of methods for implementing and following-up on WSIS initiatives;
- the political chapeau;
- the mechanisms for information society financing for development.

Subcommittee A on Internet governance met seven times in its full session in Tunis in the days immediately before the summit, while Subcommittee B (everything else) met four times after the PrepCom resumed. Consensus was reached late on the evening of November 15, 2005, approximately twelve hours prior to the official opening of the Tunis Summit.[59]

The Tragedy of Tunis: Hypocrisy over Human Rights, Civil Society's Participation Is Compromised at the Summit

The WSIS's credibility as an international event that would pave the way for an information society that would be democratic, inclusive, and respectful of human rights was seriously damaged by the demands made by the Tunisian regime on civil society activists who had been invited to participate in events running parallel to the summit.

A *Citizens' Summit on the Information Society* was organized during PrepCom III to run parallel to the WSIS in Tunis from November 16 to 18, 2005 by a civil society coalition comprised of 19 organizations. The event aimed to address a series of issues hitherto neglected in official negotiations, as well as to allow for various activists to express their positions on a number of themes related to the information society.[60]

Tunisian authorities placed a number of obstructions in the way of the civil society groups organizing and participating in the event with the aim of preventing the event from taking place. The authorities even went as far as impeding one of the meetings held on November 14, 2005, through violent means. Some members of civil society were harassed and beaten by police. The event's website was also blocked. The general difficulties experienced in organizing citizen events at Tunis were considerable. The following forms of repression were documented in relation to the WSIS:

---

59 The "Tunis Compromise" on Internet governance is discussed in Chapter 5, the follow-up and implementation mechanisms in Chapter 6 and issues associated with funding in Chapter 4.

60 See the Citizens' Summit on the Information Society website. http://citizens-Summit.org/objectives.html

- physical aggression towards foreign journalists, notably French journalist Christophe Boltanski, attacked and stabbed on the street as police watched;[61]
- Tunisian authorities' repeated surveillance of certain civil society members and foreign journalists;
- forced closing of the Citizens' Summit on the Information Society and blocking of the website dedicated to the event;[62]
- rejection from entry at the Tunisian border and forced repatriation to France of the Secretary-General of the NGO Reporters Without Borders, Robert Ménard;[63]
- repeated blocking of Internet websites before, during, and after the summit.[64]

In protest, several official events that were supposed to include civil society were cancelled by the organizations taking part in them. The regrettable situation was summed up by Ralf Bendrath:

> organizing has been difficult since PrepCom III where planning meetings in Geneva were similarly disrupted. In Tunis confirmed, pre-paid bookings for the venue were cancelled due to pressure from the authorities. As of the night of 14 November the CS website has been blocked inside Tunisia, with the exception of the media centre inside the Kram [the main Summit venue]. This solidarity action to cancel our side events on November 15 is intended to:
>
> 1. express our solidarity with the many Tunisian individuals and organizations whose basic human rights are routinely being violated
> 2. encourage all delegates at the WSIS to raise the issue of human rights violations in Tunis with their national delegations. These violations were clearly documented prior to the Summit by the IFEX Tunisia Monitoring Group (International Freedom of Expression).

---

61 See Reporters Without Borders and International Freedom of Expression eXchange, *"Libération" Correspondent Assaulted, Stabbed on Tunis Street.* (November 14, 2005). http://www.canada.ifex.org/en/content/view/full/70455. See also Reporters Without Borders and International Freedom of Expression eXchange, *Amid Worsening Pre-summit Tension, French TV Crew Pulls Out crew Because of "Close Surveillance."* (November 16, 2005). http://www.canada.ifex.org/en/content/view/full/70528

62 See the Heinrich-Böll-Foundation website for the WSIS, *The Citizens Summit is Dead – Long Live the Citizens Summit! CSIS Press Conference Becomes Major Human Rights Gathering.* (November 16, 2005). http://www.worldsummit2003.de/en/web/830.htm

63 See Reporters Without Borders and International Freedom of Expression eXchange, *RSF Secretary-General Prevented from Attending WSIS.* (November 17, 2005). http://www.canada.ifex.org/en/content/view/full/70536

64 See Reporters Without Borders and International Freedom of Expression eXchange, *IFJ Protests over Ban of its Website.* (January 5, 2006). http://www.canada.ifex.org/en/content/view/full/71383

3. be a firm reminder to everyone that the goals of the WSIS can not be achieved without respect for human rights, including freedom of expression, association and opinion, as outlined in paragraph 1 of the Geneva declaration
4. suggest that in future the United Nations gives careful consideration to hosting events of this nature in countries where the necessary preconditions for people meeting and working together peacefully do not exist.[65]

Thus, the WSIS opened in a tense climate and was irrecovably tainted by the hypocrisy of the Tunisian regime, which had, while aiming to present a modern, progressive Tunisia to the international community, resorted to repressing its own people as well as many visitors who had been officially invited to participate in the very event that was supposed to showcase this illusion.

The cancellation of the Citizens' Summit, due to the pressure and intimidation exerted by Tunisian authorities, had repercussions at the highest levels of the summit itself. A diplomatic malaise was provoked by the president of the Swiss Confederation, Samuel Schmid, when he reminded Tunisia of its duties regarding human rights using blunt diplomatic language during the opening plenary session:

> Of the many individuals who still do not have access to information resources, for many this is due to political reasons. It is not acceptable—and I say this without beating about the bush—for the United Nations Organization to continue to include among its members those States which imprison citizens for the sole reason that they have criticized their government or their authorities on the Internet or in the press. Any knowledge society respects the independence of its media as it respects human rights. I therefore expect that freedom of expression and freedom of information will constitute central themes over the course of this Summit. For myself, it goes without question that here in Tunis, within its walls and without, anyone can discuss quite freely. For us, it is one of the conditions *sine qua non* for the success of this international conference.[66]

Is This the End? The Tunis Summit, 16–18 November, 2005

International summits themselves are essentially political events where pre-negotiated agreements are officially ratified and presented to the public. These

---

65 Ralf Bendrath, *Human Rights Solidarity Action by International Civil Society Organisations: Cancellation of Several Civil Society Side Events on November 15 2005.* (November 15, 2005). http://www.worldSummit2003.de/en/web/822.htm
66 Samuel Schmid, *Second Phase of the WSIS, 16–18 November, Tunis: Statement from the President of the Swiss Confederation.* (November 16, 2005). http://www.itu.int/wsis/tunis/statements/docs/g-switzerland-opening/1.doc

global events therefore serve to raise public awareness of important social, political, economic, or cultural issues that have already been debated. The WSIS was no exception to this. Because consensus is reached during the preparatory process—even when the preparatory process concludes on the day before the summit itself—the summit events are held mostly for the purposes of public relations, photo ops, networking, and exploring future plans.

The Tunis Summit brought together members of the private sector, civil society organizations, international organizations, and government delegations in a technological trade show and conference where the different actors presented their ideas, initiatives, and products. Numerous CS events were added to official government events such as plenary meetings and talks on the different aspects of the information society. These side events encouraged discussion and dialogue on the specific themes or key issues of the WSIS.

Civil society organizations took advantage of this event to begin drafting a second phase CS declaration: *Much More Could Have Been Achieved*.[67] This document, which later became an official contribution to the Tunis output, was drafted by compiling contributions from many different civil society organizations, mostly on-line. The document is a critical evaluation of the second phase of the WSIS as well as its results. It will be discussed in the next chapter.

The two official documents of the second phase, the Tunis Commitment and the Tunis Agenda for the Information Society, were officially adopted on November 18, 2005, at the government plenary. In Part III, we will see how the end of official negotiations in Tunis in November 2005 marked the beginning of a third phase of the WSIS. This phase, which may be called the post-WSIS phase, cleared the way for establishing communication governance mechanisms that would have a significant long-term impact on the organization of existing mechanisms for communication governance.

---

67 This document is included in the appendix to the present volume. The full reference is: WSIS Civil Society Plenary, *Much More Could Have Been Achieved: Civil Society Statement on the World Summit on the Information Society*. (December 18, 2005 Revision 1—December 23, 2005). http://www.worldSummit2003.de/download_en/WSIS-CS-Summit-statement-rev1-23-12-2005-en.pdf

• CHAPTER THREE •

Civil Society at WSIS Phase II: A Summary Assessment

New Phase, New Context, New Structures?

Despite being subject to frequent criticism, the civil society internal participation and decision-making structures established during Phase I generally yielded good results. Civil society as a stakeholder was able to produce substantial and consensual collaborative documents, draft joint declarations, and establish relatively inclusive and transparent voting and participation procedures.

The period separating the end of the first phase and the first PrepCom of the second phase (December 2003–June 2004) was marked by the emergence of new controversy over the modalities of civil society participation in the WSIS. Various civil society meetings in Phase II, PrepCom I, held in Hammamet, Tunisia, were essentially hijacked by a contingent of Tunisian government-sponsored agitators determined to block any attempt on the part of WSIS CS to criticize the Tunisian regime's record on upholding human rights. The fact that, on the face of it, there were no existing procedural restrictions that could be applied to deny these Governmental Nongovernmental Organizations (GNGOs) access to civil society structures nor refute their claims to be entitled—as representatives of civil society—to have their voices reflected in WSIS CS outputs brought to light the inadequacy of some of the informal elements of the internal structures that had been used to coordinate and organize WSIS CS during the first phase. Additionally, this situation exposed a larger unease that many civil society participants had felt toward the participation and decision-making mechanisms set up during Phase I. The absence of clear rules, established working methods, and clearly defined decision-making procedures in key civil society participation processes contributed to a general sense that CS procedures lacked transparency, legitimacy, and efficiency. Over the course of Phase I, this unease had largely been left to linger below the surface because things seemed mostly to be working.

But the drama at PrepCom I underlined that reliance on such ad hoc practices might not be sustainable in the context of WSIS Phase II, much less beyond the WSIS. In particular, questions were asked about how greater membership diversity might lead to the development of new conflicts, the organizational implications of the summit being physically located in a territory governed by an authoritarian regime, and the risks associated with the increasing institutionalization of civil society participation mechanisms. What emerged, in response, was a push from within CS to review the efficiency and legitimacy of CS's existing legacy structures. Two separate review processes were launched at the beginning of Phase II that aimed to determine whether or not the structures set up for the first phase remained adequate in the political context in which civil society found itself at the onset of Phase II of the WSIS:

- a sub-group of the CSB was created at the end of Phase I to review the Bureau's mechanisms and composition and their appropriateness to Phase II;[1]
- a Working Methods Working Group was created in the aftermath of PrepCom I "to formulate proposals and recommendations on procedures, modalities and processes for civil society to work most effectively and democratically together and to serve as the collective memory for civil society."[2]

The notion of "collective memory" is important. The review of the participation mechanisms would, it was hoped, affect more than just the WSIS: members of civil society were determined to leave a legacy of organizational practices that could be used to support the inclusion of civil society in future conferences, summits and other global governance forums. In this sense, civil society participants at the WSIS were conscious of the role that the summit might play in the eventual reorganization of global governance practices, and understood the necessity of creating an organizational model that would be portable to other global policy forums.[3] The evaluation of existing CS structures would contribute to the realization of this goal not only by granting WSIS CS an opportunity to refine its model, but by allowing CS to collect

---

1 Viola Krebs, *Synthesis– CS Bureau Meeting– 26/06/04*. [WSIS CS-Plenary]. (June 26, 2004).
2 Elizabeth Carll, *RE: Charter of Working Methods Working Group*. [WSIS CS-Plenary]. (September 21, 2005).
3 See, for example, the Heinrich-Böll-Foundation website for the WSIS, *Civil Society Processes in WSIS Phase II: Adaption of Working Methods Started, Lessons Still to Be Learned*. (November 24, 2005). http://www.worldSummit2003.de/en/web/691.htm

available information and document the practices that had been developed for wider diffusion and future use.

Civil Society Review of Structures at Phase II

The tensions caused as the ad hoc and informal working methods devised for Phase I were stretched and challenged by the changing context of Phase II were evident across a network of WSIS CS structures that included:

Civil Society Bureau: the Bureau was an interesting paradox; the body was the only civil society structure to have been recognized by all WSIS actors during both phases of the summit, and was therefore a significant accomplishment and a historic first for the United Nations. Nevertheless, it remained highly controversial, particularly in regards to its internal participation mechanisms and working procedures.

Content and Themes Group: during the first phase, C & T's work mostly consisted of compiling the contributions from the various caucuses, drafting collaborative documents, encouraging the development of common positions on key WSIS issues, and translating the major texts drafted by civil society at the summit. C & T's work ended with the conclusion of Phase I and was re-started for Phase II. However, despite various calls to revive the level of dynamism that had been evident in C & T in Phase I, the "migration" of civil society participants (the arrival of new players, departure of others, as well as the movement of established groups and organizations) and the absence of the type of leadership and initiative that had been present during Phase I constrained the role of C & T during Phase II.

Plenary: the influx of new civil society participants that occurred over the course of Phase II coupled with the need to integrate often competing perspectives on the polarizing issues of Phase II, such as Internet governance and human rights in Tunisia, functioned to erode the viability of consensus-based decision-making in the Plenary. As a result, the modes of Plenary participation and decision-making had to be reevaluated.

Caucuses and working groups: the shuffling of the political agenda—in particular its shift from a broad focus on a panoply of information society issues to a more focused concern for a handful of specific issues including Internet governance and follow-up and implementation—required internal reorganization and remobilization, re-launching of activities, and creation of new groups and caucuses relevant to the second phase. The departure from the process of individuals who had been important organizers and focal points during Phase I, the emergence of a new generation of CS leadership and the influx of CS actors who had not participated in Phase I also contributed to this shuffling of the deck. In the process, questions were asked about the definition of caucuses and their membership, their linkages to each other and to other CS structures, and more generally about how some more formal parameters could be introduced to make the caucus system more transparent and legitimate.[4]

Bureau Reform

In addition to launching a sub-group of the Bureau mandated with evaluating existing Bureau organization and immediately following PrepCom I of Phase II, the members of the Civil Society Bureau announced that all "families" intending to maintain their official recognition by the Bureau were required to submit operational reports before September 15, 2004.[5] This deadline was postponed twice and finally extended to December 31, 2004. Families that did not respond to this demand had their seat in the Bureau suspended. The decision to embark on this process had three justifications: it would identify the groups that were still active following the end of Phase I, determine the roles played by different actors and groups that were active during Phase II in the structures of the Bureau, and obtain information on the activities of the groups taking part in the WSIS. The following objectives were identified for the gradual reform of the Civil Society Bureau during a Bureau meeting in Capetown, South Africa, on December 5 and 6, 2004:

> To summarise the reflections on CSB reform: although the role of the CSB has enjoyed a high reputation, especially on the governmental level, it has a lower regard by civil society in general, lacking the perception of legitimacy and clarity of its role. The aim for a structural reform of the CSB is to keep it as a useful and necessary organ to deal with the governmental process, and simultaneously to acquire greater internal civil society support, in order to ensure effective participation of civil society during the WSIS process. It was agreed that the goal of a CSB reform would be to allow a "bottom-up" approach of the different civil society constituents, by ensuring that the CSB reflects the substantive work of the CSOs during Phase II, and to correspondingly change the composition of the CSB.
>
> Participants re-affirmed the important role of the CSB as providing institutional progress in the UN / civil society relations and as a space, which needs to be filled with expertise and commitment. They also reaffirmed the Bureau as a procedural and not

---

4 For more detailed discussion of each of these structures, see Chapter 1.
5 See Meryem Marzouki, *Synthesis–CS Bureau Meeting –26/06/04.* [EN/FR] Fwd: [WSIS CS-Plenary]. (June 30, 2004).

substantive body. However, it was pointed out that it could have a political role as a facilitating body with other stakeholders.[6]

The review process concluded that it was necessary to evaluate the concept of CS "family," reorganize the participation mechanisms of the Bureau, and increase its transparency at all levels in order to make this body a legitimate tool for all members of civil society participating in the summit. The reform process for the Bureau took place over the course of the entire second phase.

The Working Methods Working Group

In the aftermath of a UNICT Task Force meeting held in Berlin in the autumn of 2004, various civil society actors met to discuss the WSIS and CS participation in the summit going forward. A wide range of questions were asked about CS working methods and structures, including:

- How did civil society operate in the first phase of the summit? What worked and what could have been improved?
- Are there some operational optional standards of transparency and legitimacy that all civil society caucuses and working groups should adhere to?
- What would be possible best working methods? What are some options for a voluntary "charter" for caucuses to adopt for their internal workings?
- How does civil society choose from its diversity speakers and representatives for particular meetings, i.e., WSIS plenary sessions, multi-stakeholder processes, press briefings, etc.?[7]

On this basis, a group of CS participants[8] launched the Working Methods Working Group, confirming and channeling the groundswell desire that the internal participation structures for civil society at the WSIS be reformed.

> The Working Methods Working Group is intended to bring together civil society actors to work together to formulate proposals and recommendations on procedures, modalities and processes for civil society to work most effectively and democratically

---

6 Civil Society Bureau, *Civil Society Bureau Meeting, Cape Town, South Africa, 5–6 December 2004: Summary.* (December 15, 2004). http://www.un-ngls.org/wsis%20CSB%20-%20Cape%20Town%20-%20Summary%20and%20Outcome%20-%20FINAL.doc
7 Vittorio Bertola, Christoph Bruch, Jeanette Hoffmann, Ramin Kaweh, Nnenna Nwakanma, and Rik Panganiban, *Proposal for Formation of CS Working Methods Working Group.* (November 22, 2004). http://www.worldsummit2003.de/en/web/690.htm
8 The following individuals launched the Working Methods Working Group initiative: Vittorio Bertola (ICANN At Large Advisory Committee), Christoph Bruch (Humanistische Union), Jeanette Hofmann (Internet Governance Caucus Coordinator), Ramin Kaweh (UN Non-governmental Liaison Service), Nnenna Nwakanma (African Civil Society for the Information Society), Rik Panganiban (Conference of NGOs in Consultative Relationship with the United Nations).

together. These recommendations are intended for use by other civil society bodies in the WSIS process, including the Civil Society Bureau, Content and Themes Group, Caucuses, Working Groups and the Plenary writ large. Membership in the working group is open to all interested civil society actors.[9]

In contrast to the Bureau review process, this review would comprehensively address all civil society structures and draw its membership bottom-up. Pointedly, reform of the Bureau itself was included as agenda item one in the list of issues to be addressed, in spite of the WMWG's founders being entirely aware of the Bureau's own internal review that was proceeding in parallel.

The proposed organizational audit was welcomed by many CS participants, but the risks associated with it were also acknowledged. Ralf Bendrath pointed out that civil society had to establish new working and participation methods that would reinforce its inclusive nature and its legitimacy while avoiding feeling obligated to impose a bureaucratically rigid and fixed structure:

> The overall procedural challenge is to develop working methods and principles that increase civil society legitimacy and inclusiveness, and at the same time avoid a too bureaucratic structure. The latter would in the end either been followed by nobody or it would deprive civil society of its biggest advance – the ability to stay flexible and act on very short notice. And the working methods must not prevent individual groups or networks from being able to raise their authentic voice and their clear concerns; otherwise they will just work on their own again.
>
> The strategic but not smaller challenge is more political and less procedural. Civil society has to avoid being drawn too much into the official process and resist the temptation to replicate the intergovernmental structures. Otherwise, it will end up with a group of professional NGOs that are recognized by the governments, belong to the international conference and policy jet-set, even might have some influence here and there, but are more or less decoupled from the grassroots work and the more radical positions of the broader social movements. As a participant at the Berlin meeting stated, "multi-stakeholder processes are enabling and including, but also disciplining."[10]

The pace of discussions and actions around the work of the WMWG was plodding. It took the better part of a year to review existing structures and then devise and adopt guidelines for civil society modalities of participation. The roadmap proposed by the WMWG included reforms to:

---

9   Rik Panganiban, *Information Note on Working Methods Working Group at PrepCom II*. [WSIS CS-Plenary]. (February 22, 2005).

10   Ralf Bendrath, *Civil Society Processes in WSIS Phase II: Adoption of Working Methods Started, Lessons Still to Be Learned.* (November 25, 2004). http://www.worldSummit2003.de/en/web/691.htm

- **CS Plenary chairing**: The group suggested that the main purpose of the CS Plenary is for information exchange and reporting on working groups. The role of the chair is therefore to be a "facilitator" of the information exchanges between the caucuses and working groups, as well as the monitoring reports from the intergovernmental negotiations, and has no political role. Based on this, the WMWG suggests rotation by regions or direct election from plenary as two possibilities for selecting the chairs of the CS plenary sessions. It recommended to select two individuals who would know about the processes and CS organization in general, have facilitating skills, and not be reporting to the Plenary themselves.

- **Content and Themes re-structuring**: WMWG suggests a fuller "charter" of content and themes be developed, including the general scope of the CT Group and the decision-making procedures. In principle Content and Themes should be open to wide participation of all civil society. One of its main tasks is to decide, on a consensual basis, on those caucuses speaking at the civil society interventions during the PrepComs.

- **Mediation Group for Civil Society disputes**: The WMWG agreed that there were both procedural and political matters that led groups to have disputes with each other, both of which were important to address. WMWG agreed with the suggestion put forward to it, that the Bureau should explore establishing a group of mediators available for any mediation requests from civil society groups.

- **Caucuses and Working Groups' structure, procedures, membership issues, etc.**: WMWG notes that a caucus of civil society exists to enable civil society organizations and their representatives to share information and views, coordinate activities among members, draft agreed caucus text, and select speakers for official plenaries, press conferences and other meetings. WMWG emphasizes importance of the principle of openness. Meetings of caucuses should be open to all groups, with some fair criteria on membership in the caucus.

    Those creating new caucuses should announce their intention publicly both at a physical plenary meeting and the plenary email list. It is recommended that all caucuses should have a charter including their mission, activities, structure, composition, and procedures, in accordance with general principles. Caucuses should have available a contact point, a list of their members, at least one open meeting at WSIS PrepCom meetings, an email list and an open archive.[11]

Many of the WMWG's recommendations were eventually adopted by the Civil Society Plenary at the end of September 2005, when the summit was in the midst of the third and final PrepCom of Phase II. Despite this questionable timing, the effort may prove to be historically relevant as it addresses a series of fundamental questions regarding the inclusion of civil society in global politics.

---

11 Rik Panganiban, *Information Note on Working Methods Working Group at PrepCom II*. [WSIS CS-Plenary]. (February 24, 2005).

Table 2: Definitions of and Guidelines for CS Structures

Civil Society Plenary Created at PrepCom 1, July 2002	**1. Mission** "The Mission of the WSIS Civil Society Plenary (CSP) is to bring all Civil Society Organizations (CSOs) and individuals together for information exchange and reporting; in special cases, a decision-making CSP meeting can be organized for strategic, procedural and general civil society related decision-making purposes." **2. Objectives** • to provide information exchange and reporting by the different CS structures; • to foster global civil society decision-making; • to encourage a sense of global civil society community for CS actors participating at WSIS. **3. Goals** • to provide a space for dialogue and information exchange; • to enhance the coordination and effectiveness of CS structures in the WSIS; • to develop consensual positions on specific issues where possible.
Content and Themes Working Group Created at PrepCom 1, July 2002	**1. Mission** *"The Content and Themes Group (CTG) is responsible for all matters relating to content, issues, and themes pertaining to the WSIS."* **2. Objectives** • to coordinate the work of Caucuses/Working Groups and other content related entities; • to facilitate discussion and agreement and take decisions on content-related issues; • to organize and to coordinate texts on any content-related matter/issue coming from the CTG in the name of WSIS civil society. **3. Goals** • to provide a forum for discussing overall CS strategy and informal advocacy activities; • to coordinate the compiling of speaking slots among Caucuses/Working Groups for the official intergovernmental meetings; • to delegate and organize the drafting of texts; • to organize content-related events, such as press briefings; • to report regularly to the CS plenary on decisions taken;
Caucuses and Working Groups Created at PrepCom 2, Feb. 2003	• Discuss relevant issues between members; • Gather and exchange information on content and participation issues between members; • Draft collective documents on issues to send to other civil society organizations and groups, as well as to WSIS delegates and instances; • Caucuses and Working Groups may participate as members of C& T if they can satisfy these conditions: 1. having a statement of intent; 2. a contact point and (partial) list of members; 3. at least one open meeting, preferably more, at the current WSIS-related event; 4. a Discussion List and open Archive.[12]

Continued on following page

---

12 With the exception of elements of the section entitled "caucuses and working group," the information in this chart was taken from the orientation document distributed by the NGO CONGO during PrepCom III of Phase II. See Conference of Non-Governmental

Tabl;e 2 continued

Civil Society Bureau Created at Prep-Com 2, February 2003	The Civil Society Bureau (CSB) is the interface between civil society and intergovernmental participants in WSIS. It also interacts with the Executive Secretariat and other stakeholders on procedural issues. The CSB works in parallel and interactively with the Intergovernmental Bureau. It is a mechanism that facilitates interaction among actors, thus fostering a more concrete and effective approach to multi-stakeholder negotiation processes. The CSB has an operational role, concerning itself with logistical needs, procedures and interactions. It does not deal with content or substantive issues, but it aims to facilitate the work of civil society in the process and to enhance effective partnership and interaction with other stakeholders. The CSB reports to and communicates with CS Plenary on logistical matters, and consults with CS Plenary on logistical matters having content implications. Mission • To facilitate the effective participation of civil society in the WSIS process; • To enhance effective partnership and interaction among civil society and other stakeholders; • To deal/set/establish and to facilitate procedures of, and interactions between, working groups and families • To be a communication channel for the formal working processes of the Summit Tasks; • Procedural advocacy to ensure effective civil society participation in all possible WSIS processes; • Logistical facilitation of meeting spaces, resources, interpretation, liaison, translation and scheduling; • Processes to organize and resource civil society meetings Composition. The CSB comprises focal points of groupings/families from civil society, ensuring that it reflects WSIS agenda. Its composition is regularly reviewed. Non-active groups are removed and new working groups or returning groups are enabled to join the CSB

WSIS CS v. 2.0: The Evaluation/Reform Outcomes

CS could have opted for significant reforms to the existing model or even for a complete redesign of its structures. It was determined, however, that the existing structures should largely endure, subject to some reforms aimed at increasing their transparency and efficiency but avoiding any actions that might weaken civil society's capacity for cooperative organization at the WSIS. In particular, efforts were made to formalize the guidelines and role definitions

---

Organizations in Consultative Relationship with the United Nations (CONGO), *Civil Society Orientation Kit*. (November 2005). http://www.ngocongo.org/congo/files/wsis_ oriention_kit.pdf

outlined in the above chart and have them recognized and approved by WSIS CS.

In September 2005, the Civil Society Plenary approved the detailed mandates of each structure, which included their internal regulations and operational procedures. This information was diffused as part of an orientation kit that was distributed to both new and established civil society members at PrepCom III and then again at the Tunis Summit.[13] Thus, reform was a partial victory for WSIS CS as more legitimate working and participation procedures were developed. The late timing of their adoption, however, considerably diminished their impact on the participation of CS within the summit itself.

In the meantime, the Conference of Non-Governmental Organizations in Consultative Relationship with the United Nations (CONGO) produced and distributed a code of conduct for the participation of non-governmental organizations in UN conferences. Despite its unofficial status, the code laid out participatory guidelines for members of civil society taking part in subsequent international forums.[14]

The internal organization processes used to organize and coordinate civil society participation at the WSIS will undoubtedly remain one of the more significant legacies of the summit in the long term. A decidely political negotiation, the WSIS was also a testing ground for the further integration of non-governmental organizations into the UN system. The ability of civil society groups to organize themselves in a relatively inclusive and transparent manner despite all of the controversy, agitation, and deep ideological and political disagreements surrounding their participation proved to government stakeholders that it was possible to maintain a coherent and organized dialogue with global civil society as a credible, official conversation partner.

CS Refuses to Remove Its Foot from the Doorway: The Campaign for Multi-stakeholder Global Governance at Phase II

As we have seen, CS substantive contributions to Phase II were defined by two trends: the broad WSIS political agenda of the first phase narrowed to focus on only a handful of issues in the second phase and the composition of

---

13 Conference of Non-Governmental Organizations in Consultative Relationship with the United Nations (CONGO), *Civil Society Orientation Kit*. (November 2005). http://www.ngocongo.org/congo/files/wsis_oriention_kit.pdf
14 Ibid.

the CS contingent participating in the WSIS as well as the polarizing tenor of debate within CS over certain issues meant that consensus on substance occurred with less frequency. Participation became a logical preoccupation of certain segments of CS at Phase II. It was a cross-cutting issue that could engage CS participants not directly involved in the Internet governance or follow-up and implementation debates. Furthermore, threats to continued or further CS participation represented logical rallying points for bringing together all sectors of CS and thus provided a rare example of a Phase II issue on which a consensusal CS viewpoint could be expected.

Thus, over the course of the second phase of the WSIS, a constant and galvanizing focus of WSIS CS activity was the effort to assure, protect and expand on the participation rights that had been gained by CS during Phase I. The agenda for CS's Phase II participation campaign was set and communicated to other WSIS stakeholders at PrepCom I. In a speech delivered by Ralf Bendrath on behalf of the Civil Society Plenary, CS insisted that governments had, by virtue of the precedents set at Phase I, already acknowledged that "governments can not address these issues alone" and that "any mechanism that does not closely associate civil society and other stakeholders is not only unacceptable in principle, it is also doomed to fail." Bendrath demanded that "the multi-stakeholder process be treated not just as "a nice phrase," but "becomes true reality." Going as far as threatening that CS's "further participation" in the WSIS would be dependent upon these conditions being met, Bendrath concluded with a sternly worded warning that "we are not willing to play an alibi role or lend our legitimacy to a process that excludes us from true meaningful participation. The summit can only be a summit of successes if there is substantive progress in our participation."[15] These prospects were quickly undermined.

A Major Step Back: The Group of Friends of the Chair

The first meeting of the preparatory committee for Phase II introduced a new working group called the Group of Friends of the Chair that was given the task of preparing documents intended to be the basis for negotiation at the next preparatory committee meeting. As we chronicled in Chapter 2, the work of the GFC was conducted throughout the entire preparatory process leading up to Tunis. Its main task was to define the general structure of the texts to be negotiated over the course of Phase II. The role of proposing agendas and

---

[15] This speech is discussed in further detail in Chapter 2. See Ralf Bendrath, *Statement to the PrepCom Plenary on Behalf of the Civil Society Plenary.* (June 25, 2004). http://www.itu.int/wsis/docs2/pc1/plenary/heinrich-boll.doc

compiling propositions was a decisive one, and the group entrusted with these tasks had the potential to greatly influence summit negotiations. This group had the ability to take initiative in its first task, which was to propose a vision and structure for the documents that were later to be negotiated—i.e., a political vision for the information society (the Tunis Commitment), and an operational plan that aimed at transforming this vision into concrete measures (the Tunis Agenda for the Information Society). The content, style, and structure of these final documents would of course still be reviewed and approved by the full government plenary. Yet, the GFC proposed working methods, negotiation agendas, and the general structure for the documents to be adopted, thus playing a strategic role of great importance while fulfilling its mandate to ensure the smooth flow of negotiations and compiling the positions of different actors.

Though PrepCom President Karklins had introduced the GFC structure in the hope that it could be formed with a multi-stakeholder composition, certain government delegations seem to have been able to successfully block those efforts.[16] Thus, despite this clear agenda-setting role, the make-up of the GFC was strictly intergovernmental, restricting non-state actors to observer status at its open meetings.[17] This loss of influence over the political direction of the summit was seen as a significant setback to the multi-stakeholder model and a serious retreat from the level of influence over the agenda achieved by CS at WSIS Phase I.[18]

From the perspective of CS participation, the results achieved in the other working groups convened during Phase II could not have been much worse than the GFC experience. They were, however, decidedly mixed.

The Strategic Dimension of Expert Committees: The WGIG and the TFFM

The technical and regulatory complexity of the issues discussed at the WSIS was considerable. These issues were so complex that even the task of arriving at common definitions emerged as an occasionally difficult obstacle in the negotiation process. Many government delegations simply lacked access to the

---

16   Heinrich-Böll-Foundation website for the WSIS, *Drafting Process for Tunis Summit Declaration is Starting: "Group of Friends of the Chair" Met Today, Civil Society Mostly Locked Out.* (October 21, 2004). http://www.worldSummit2003.de/en/web/677.htm

17   President of the PrepCom of the Tunis Phase, *Report of the Work of the Group of the Friends of the Chair During the Inter-sessional Period.* (WSIS-II/PC-3/DOC/6-E). (September 8, 2005). http://www.itu.int/wsis/docs2/pc3/html/off6/index.html

18   Heinrich-Böll-Foundation website for the WSIS, *Drafting Process for Tunis Summit Declaration is Starting: "Group of Friends of the Chair" Met Today, Civil Society Mostly Locked Out.* (October 21, 2004). http://www.worldSummit2003.de/en/web/677.htm

sort of specialist knowledge required to meaningfully respond to some of the issues being negotiated. This was particularly true of the ongoing disagreements over two fundamental issues—Internet governance and the financing of the information society. In response, the first phase ended in an agreement that called for working groups to be convened to study and draft reports framing government negotiations and clarifying misunderstandings and knowledge gaps around how these issues were being discussed at WSIS. The mandates of the two committees were noticeably different. While the TFFM was limited to the task of evaluating the efficiency of current financing mechanisms and preparing a report on the political debate surrounding it, the WGIG was given the task of proposing policies for Internet governance.

Civil society was quick to recognize the importance of participating in these expert committees. As official working groups, mandated by the United Nations to produce reports to assist or frame government negotiations, they presented compelling opportunities for leveraging CS's specialist knowledge and practical experience as a stakeholder group in order to wield political influence and help set the agenda for political negotiations at Phase II of the summit. As such, it was in the best interest of civil society groups to push for the broadest possible access to and participation rights within these committees. Considerable effort was directed towards this goal, with mixed results.

At the first public consultation session, held in Geneva on the 20th and 21st of September, 2004, the Civil Society Internet Governance Caucus insisted that, in terms of its composition, structure, and operating guidelines, the WGIG must:

- remain independent of WSIS Preparatory Committee Meetings (PrepComs);
- be constituted at the working level rather than as a "High Level Group";
- include regional meetings in its consultation process in order to provide greater opportunities for input from civil society and other entities;
- be composed of a membership balanced equally between participants from governments, the private sector and civil society, not favoring one group over any other.[19]

Those charged with setting up the WGIG seem to have largely followed these recommendations. The WGIG turned out to be the most open, inclusive, and transparent site for multi-stakeholder activity at the WSIS. The group

---

19 WSIS Civil Society Internet Governance Caucus, *Contribution to the Working Group on Internet Governance (WGIG)*, First Open Consultation 20-21 September, 2004. (August 29, 2004). www.un-ngls.org/orf/csig-caucus.doc

was ultimately composed of government representatives, members of the private sector, and civil society organizations, thus reflecting the multi-stakeholder principle in a very equitable way. The methods for participation within the group were entirely open to civil society, as working group members participated as equals, regardless of their constituency. This was a marked difference from usual WSIS practice, where—even at the best of times—CS and private stakeholders deferred unequivocally to government delegations on the contents of official documents. CS members of the WGIG were able to meaningfully influence the discussions within the WGIG, impose their own views, communicate the perspectives of other civil society actors and, perhaps most fundamentally, build trust and personal working relationships with key individual government delegates. The WGIG experience and the extent to which CS perspectives were able to influence the IG debate that followed it were unequivocally seen as an important victory for civil society organizations and as the instance during the WSIS when the multi-stakeholder experiment was pushed to its furthest and best able to demonstrate its full potential.[20]

Civil society participation in the Task Force on Financial Mechanisms (TFFM) was smaller-scale and could be considered the cynical antithesis of the full multi-stakeholder principle that the WGIG reflected so well. The TFFM was officially launched on October 4, 2004, and was placed under the supervision of UNSG Kofi Annan. The United Nations Development Programme (UNDP) was given the organizational lead. The mandate assigned to the TFFM was criticized by civil society organizations; its scope was limited to analyzing and evaluating the efficiency of the current financial mechanisms available to populations marginalized from global communication networks. Furthermore, the TFFM avoided undertaking any new empirical or theoretical research by relying on studies and reports that had already been conducted and were available at the time of the group's inception.[21]

Although the 24-member TFFM was established in October 2004, some of the work seemed to have been undertaken by the UNDP prior to that date. Civil society organizations saw the member selection and nomination processes of the TFFM as opaque and top-down, in stark contrast to those set up

---

20   For an overview of the WGIG report and its membership see the WGIG website http://www.wgig.org/index.html. For more detailed desciptions of and reflections on the WGIG experience written from the perspective of its membership, see the collection edited by William Drake (ed.), *Reforming Internet Governance; Perspectives from the Working Group on Internet Governance*. New York: UN ICT Task Force. 2005.

21   Chantal Peyer, *Geneva Informal Meeting on Financial Mechanisms Disappointing: But View of Information and Communication as "Global Public Good" Gains Support*. (November 26, 2004). http://www.worldSummit2003.de/en/web/697.htm

by the WGIG. The TFFM submitted its final report in December 2004 and its work was a basis for negotiations that took place on related issues during PrepCom II in February 2005.

Civil society was extremely critical of the drafting process of the final report, of the results achieved through the entire TFFM process and in particular, the group's lack of transparency, inclusion, and openness. The composition of the TFFM, though nominally multi-stakeholder, included only a small CS contingent.[22]

The similarities and differences between the WGIG and the TFFM are significant. The Geneva Plan of Action had mandated both these bodies to gather and organize the expertise that would frame government negotiations in the effort to focus and support particularly arduous negotiations. The WGIG was given more or less a year to complete its work. The highly technical and controversial nature of its subject called for a larger group of stakeholders to be involved so that a variety of backgrounds, experiences, and specialist knowledge bases would be covered. Procedural matters were emphasized at WGIG. The success and legitimacy of its work was seen to hinge on its levels of transparency, participation, and inclusion. Candidate nomination procedures were, to a degree, bottom-up, and different stakeholder groups were permitted to nominate the individuals they deemed to be competent and appropriate. Government delegates, the private sector, and civil society were able to closely follow the developments and activities of the WGIG through its formal consultation and interim report process as well as through the informal communication flows between members and stakeholder groups. The final WGIG report included a broad range of policy recommendations.

As for the TFFM, the official deadline that was unreasonably given was slightly more than two months. The task of managing the work was given to the UNDP, a UN agency specialized in the same field as the work facing the task force. The UNDP already had its organizational approach to the issues involved, its own experts and expert networks as well as established working guidelines. The extremely short deadline created pressure and significantly impacted the ability—and even will—of the TFFM to devise meaningful, open-ended consultation mechanisms as part of its program of work. The membership of the TFFM was nominated by a top-down process that involved little external consultation. Civil society felt marginalized by the process, unable to effectively get its voices heard and opinions considered by the other members and largely without influence on the task force's work. In particular, CS mem-

---

22 See Task Force on Financial Mechanisms for ICT for Development, *Financing ICTD–A Review of Trends and an Analysis of Gaps and Promising Practices*. (December 22, 2004). www.itu.int/**wsis**/tffm/final-report.pdf

bers were alienated by the fact that the work of the TFFM was limited to revision of existing mechanisms and avoided making policy recommendations. This approach effectively precluded discussion of the sorts of fundamental alternatives that civil society typically advocates and, thus, was seen to have basically marginalized CS participation from the start.

WSIS Inclusivity in Question

By the time the working groups had wound down prior to PrepCom III, CS had experienced both the best (WGIG) and the worst (GFC, TFFM) case interpretations of the WSIS multi-stakeholder principles. Enthused by the WGIG experience, emboldened by the pending adoption of the WMWG reforms and concerned about which interpretation of the multi-stakeholder model would emerge from the final PrepCom as both the legacy of the WSIS and the precedent for future events, CS was determined to maximize its participation rights in PrepCom III.[23]

As a baseline, CS demanded direct, real-time, uncensored, and exhaustive access to all PrepCom activities where the positions of other participating stakeholders would be presented and debated. Such access was deemed to be imperative for identifying potential allies and possible adversaries, understanding the changing negotiation dynamics as well as the openings and opportunities to be seized, and communicating the positions, perspectives, and comments of civil society organizations to government delegations. The influence of civil society organizations therefore rested on their ability to reach appropriate delegates, in the appropriate places, at the appropriate time, making access to key negotiation sites a priority as CS endeavored to influence partners with considerably more resources at their disposal.

The WSIS rules regarding stakeholder participation in PrepComs had been developed and negotiated on the fly, largely in response to previous PrepComs that had used a full plenary structure. Negotiations at PrepCom III, as we have discussed in Chapter 2, were structured around two subcommittees. Furthermore, these subcommittees then often created smaller breakout drafting groups. These sub-plenary structures would come to be seen as grey zones in regard to the application of multi-stakeholderism.

---

23 In addition, these stakes were arguably only raised on the first day of PrepCom III when CS participation—in the guise of the fight over the non-accreditation of the NGO Human Rights in China—was subject to a highly political debate between governments (see Chapter 2).

There was dissatisfaction from both CS and the private sector with their respective roles in the initial stages of Subcommittee A. A chair's paper[24] was produced that the private sector Coordinating Committee of Business Interlocutors (CCBI) cited as failing to reflect stakeholder input. During the intense, charged and plodding government negotiation sessions held in full Subcommittee A plenary, the chair and secretariat had to occasionally be reminded to allow for the scheduled speaking slots reserved for CS and other observers or wrapped up sessions entirely without calling on them. However, controversy over participation truly mobilized CS around the end of the first week of PrepCom III when break-out drafting groups were convened.

The PrepCom rules were entirely unclear about what role non-governmental stakeholders had the right to demand and were permitted to play in such sessions. Certain governments insisted that these sessions—where small groups of countries representing competing viewpoints gathered to hammer out, word-by-word, text for the final documents that might be agreeable to all—were not for discussion or policy development but were strictly negotiation meetings. With negotiation of agreements at the WSIS being the right and responsibility of governments alone, these particular governments argued that such meetings should neither require nor tolerate stakeholder involvement, intervention or observation. Civil society countered that it could not be asked to add a measure of transparency to these negotiations if governments were going to actually write the text of the documents behind closed doors and in secret, that this was a major step back from the level of participation granted to CS during the closing stages of the Geneva phase, and that CS expert knowledge could help facilitate negotiations in these groups as well.

After consultation between the governments favoring that they be closed completely (Brazil, Iran, China, etc.), those promoting the idea that stakeholders be allowed in as observers only with limited speaking rights (Singapore, El Salvador, etc.), and those advocating full participation in drafting group negotiations (US, most strongly), it was initially agreed that stakeholders would be permitted to attend at the outset and present statements. From there, however, there remained unresolved disagreement between delegations insisting that stakeholders should be then asked to leave so that governments could engage in negotiations alone and those in favour of allowing stakeholders to remain in the room to observe with no further right to speak. Because the governments could not come to a consensus resolution amongst themselves, the issue was forwarded to PrepCom President Karklins.

---

24 Chair of the Sub-Committee A (Internet Governance), *Chapter Three: Internet Governance—Chair's discussion paper* (WSIS-II/PC-3/DT/8). (September 22, 2005). http://www.itu.int/wsis/docs2/pc3/working/dt8.doc

CS delegates who met with Karklins for his thoughts on the subject reported to Content and Themes that, though sympathetic, he had conceded the existence of counter-indicative interpretations of the existing rules but expressed a reluctance to himself make an ad hoc, one-time ruling on the application of these rules to drafting groups, in fear that it might establish a precedent for subsequent UN events. Officially Karklins was considering his ruling on this fraught procedural question, but in practice the situation—and the accompanying ambiguity—was left to linger.

As a result of this uncertainty, CS delegates had different experiences trying to participate in drafting groups, some were asked to "talk and walk," others were pressured to stay away or permitted to observe. One CS delegate reported that he had managed to finagle his way into the middle of the negotiations.[25] Even in the most severe cases of exclusion, however, CS delegates reported that friendly governments were more than happy to debrief excluded stakeholders at the conclusion of meetings. When pressed for clarification of exactly what the rules were by IGC co-coordinator Adam Peake, the official response from summit organizers—given with a wink and a nudge—was "we have injected constructive ambiguity in this process" and that CS should try to make the most of the uncertainty.[26]

Regardless, the drafting groups' situation had kicked CS mobilization around the issue of multi-stakeholder participation into high gear and a scathing critique was drafted, approved and delivered to a meeting of Subcommittee B. Presented by Avri Doria, the statement laid out that

> The decision to exclude non-governmental stakeholders from the drafting groups is not about rules and procedure—it is a matter of political courage and principle. You have a choice to be inclusive or exclusive, to work in partnership, transparency and openness. There is a great opportunity here to move forward with all the progress we have made within the UN and WSIS, and this move will be a move backwards. [...] We strongly protest your decision to exclude non-governmental observers from the drafting groups. Civil Society should be able to make statements on the same basis as we do in Subcommittee, to remain in the room as observers for the entire session and to further contribute at the discretion of the chair.[27]

---

25 Ralf Bendrath, *Report from Drafting Group III, Subcom A*. [WSIS CS-Plenary]. (September 27, 2005).

26 Jeremy Shtern, *Subcommittee A Notes: Plenary Sept 27 (first half)*. [WSIS CS-Plenary]. (September 27, 2005).

27 Heinrich-Böll-Foundation website for the WSIS, *Civil Society Statement on the Decision to Exclude Non-governmental Stakeholders from Drafting Groups*. (September 28, 2005). http://www.worldSummit2003.de/en/web/788.htm

In parallel, discussions were under way about organizing a possible CS withdrawal from the WSIS process altogether. This never came about but, given the extent to which CS struggled to develop consensus and integrate different opinions and elements over the course of most of Phase II, the sense of exclusion, of participation rights being rolled back, and of responsibility to create a post-WSIS legacy for CS in other venues must be seen to have been a catalytic force without which, it is unlikely, CS would have come together between PrepCom III and the Tunis Summit to draft a second CS parallel declaration (see Chapter 2). Furthermore, this campaign for full multi-stakeholder global governance mobilized and accelerated during the later stages of WSIS Phase II has continued through the convening of the Internet Governance Forum, and has expanded beyond the WSIS IGF cluster to a series of international organizations. The evolution of the campaign for CS participation and its gains are presented in Part Three of this book. In Chapter 7, we also critically reflect on what might have been lost in the move to organize CS around issues of process rather than more normative issues of substance. Before doing so, however, it is worth evaluating what exactly the WSIS Phase II did accomplish in a substantive sense. We do so in Part Two.

PART TWO
WSIS Phase II Issues and Outcomes

The official conclusion of the WSIS on November 18, 2005, marked the end of an expensive, innovative, and particularly long political process that had begun nearly eight years earlier. The official integration of non-governmental partners into the summit as well as its segmentation into two distinct phases greatly enriched the WSIS and made it more complex.

Evaluations of the results submitted to the WSIS differed significantly depending on who was doing the assessment. Not surprisingly, the organizers of the summit considered the initiative to be "a resounding success."[1] Official figures show that the following parties participated in the Tunis Summit:

- 46 Heads of State and Government, Crown Princes, and Vice-Presidents and 197 Ministers/Vice Ministers and Deputy Ministers
- 5857 participants representing 174 states and the European Community
- 1508 participants representing 92 international organizations
- 6241 participants representing 606 NGOs and civil society entities
- 4816 participants representing 226 business sector entities
- 1222 accredited journalists from 642 media organizations of which 979 were onsite from TV, radio, print, and online media worldwide.[2]

Whether the organizers of the WSIS will admit it or not, the event was only partially successful in attracting the attention of the heads of state from developed countries, as most of them were content with merely sending representatives from an appropriate ministry to Tunis. Most of the high-level figures who attended the event came from countries and regions of the world that have been adversely affected by the digital divide. Very few high-level delegations from the developed North attended the summit, which implicitly dem-

---

1   WSIS official website, *Press Release: World Summit on the Information Society Hailed as Resounding Success: Consensus and Commitment in Tunis Paves the Way to a More Equitable Information Society.* (November 18, 2005). http://www.itu.int/wsis/newsroom/press_releases/wsis/2005/18nov.html
2   Ibid.

onstrated the lack of enthusiasm from countries that did not stand to gain much from a political event dedicated to eradicating the digital divide.

However, despite this lack of enthusiasm on the part of the North and the controversies associated with the three great challenges of the second phase, namely information society financing, Internet governance, and the follow-up and implementation of adopted decisions, the political process was indeed successful in the sense that some agreement was reached. Nevertheless, the question must be asked: can this political consensus truly contribute to the eradication of the digital divide, connect marginalized populations to networks, and integrate networked information and communication technologies (NICTs) into an international development framework?

The second part of this book will present the decisions that the WSIS adopted regarding information society financing, Internet governance, and the implementation and follow-up of WSIS resolutions. It will also present the views of civil society on these questions and offer a critical analysis of the WSIS's processes, results, and legacy for the global governance of communication.

• CHAPTER FOUR •

Digital Solidarity? Financing Access to the Information Society

Rhetoric about the information society can mask a global context marked by profound inequalities. As noted in Chapter 1, according to UNESCO, more than 774 million adults worldwide are thought to be illiterate. Of that total, 64% are estimated to be women.[1] The rates of illiteracy among adults in West and South Asia as well as sub-Saharan Africa range between 30% and 40%.[2] Similarly, according to the International Energy Agency, more than 1.6 billion people lived without electricity in 2002.[3] Many populations, especially in Africa and Asia, do not have the resources necessary to acquire and maintain the infrastructure that is essential to operating modern communication networks. This disparity in resource allocation creates digital divides between regions, but also within countries, cities and even communities.

Before moving on to discuss the debates over financial mechanisms for closing the digital divide that occurred during the WSIS, it is important to situate this discussion through a quick overview of some of the issues that are related to financing for information and communication technologies for development (ICTD) and international cooperation.

---

1   UNESCO Institute for Statistics, *According to the Most Recent UIS Data, There Are an Estimated 774 Million Illiterate Adults in the World, About 64% of Whom Are Women.* Updated on October 8, 2009. http://www.uis.unesco.org/ev_en.php?ID=6401_201&ID2=DO_TOPIC
2   UNESCO Institute for Statistics website, *National Literacy Rates for Youths (15–24) and Young Adults (15+).* http://stats.uis.unesco.org/unesco/ReportFolders/ReportFolders.aspx?IF_ActivePath=P,55&IF_Language=eng
3   International Energy Agency, *World Energy Outlook 2004: Executive Summary.* http://www.iea.org/textbase/npsum/WEO2004SUM.pdf

Financing at WSIS II: Issues and Controversies

From its outset, the WSIS was developed in conjunction with the UN Millennium Development Goals. Resolution A/RES/56/183, which established the WSIS, opens with official recognition of the place of NICTs within the United Nations development objectives

> Recognizing the urgent need to harness the potential of knowledge and technology for promoting the goals of the United Nations Millennium Declaration and to find effective and innovative ways to put this potential at the service of development for all,
>
> Recognizing also the pivotal role of the United Nations system in promoting development, in particular with respect to access to and transfer of technology, especially information and communication technologies and services, inter alia, through partnerships with all relevant stakeholders.[4]

The Millennium Development Goals were adopted by the UN in September 2000 with the conclusion of the Millennium Summit. They spell out a series of objectives to be achieved by 2015 as well as the overarching target of significantly reducing extreme poverty. This initiative was accompanied by a series of indicators that measure strategic aspects of the international fight against extreme poverty and a commitment to closely monitor the situation and the degree of progress made. The Millennium Development Goals specifically target the following:

- the struggle against poverty and hunger;
- universal education;
- gender equality;
- reducing infant mortality;
- improving the health of mothers;
- the struggle against HIV/AIDS;
- preserving the environment;
- developing global partnerships on development.

The Millennium Declaration is most relevant to the global governance of communication where it connects poverty reduction directly with universal access to information and communication technology, articulating the need to:

---

4   United Nations General Assembly, *Resolution A/56/558/Add.3. World Summit on the Information Society.* (January 31, 2002). http://www.itu.int/wsis/docs/background/resolutions/56_183_unga_2002.pdf

ensure that the benefits of new technologies, especially information and communication technologies, in conformity with recommendations contained in the ECOSOC 2000 Ministerial Declaration are available to all.[5]

Thus, the more equal distribution of the benefits derived from new information and communication technologies is a fundamental goal of the Millennium Declaration, and, as such, a fundamental development objective for the 21st century. The integration of information and communication technologies into disadvantaged sectors can be further interpretated as a *cross-cutting issue* in development because these technologies allow for the resolution of social and economic issues that are only indirectly linked to technological factors. For instance, access to advanced communication systems capable of transmitting high resolution video conferencing and medical imagery allows doctors to diagnose patients several hundred kilometers away from health centre facilities. The development of such systems and their integration into local practices would therefore have knock-on benefits that would contribute to meeting other poverty reduction goals related to public health, general health, and infant mortality. In other domains, increased availability and ability to make meaningful use of advanced ICTs could have similar beneficial effects on education and literacy, employment, and economic, cultural, and political development. From a development standpoint, the integration of new information and communication technologies is important to the extent that it can open new development opportunities and encourage the creation of new social and cultural dynamics that are conducive to economic development and the achievement of a variety of development goals. Political enthsuiasm for this "digital opportunity" was defined by the following list of objectives to be reached by 2015 that was included in the Geneva Plan of Action:

- to connect villages with ICTs and establish community access points;
- to connect universities, colleges, secondary schools and primary schools with ICTs;
- to connect scientific and research centers with ICTs;
- to connect public libraries, cultural centers, museums, post offices and archives with ICTs;
- to connect health centers and hospitals with ICTs;
- to connect all local and central government departments and establish websites and email addresses;
- to adapt all primary and secondary school curricula to meet the challenges of the Information Society, taking into account national circumstances;
- to ensure that all of the world's population have access to television and radio services;

---

5   United Nations General Assembly, *Resolution 55/2. United Nations Millennium Declaration.* (September 18, 2000). http://www.un.org/millennium/declaration/ares552e.pdf

- to encourage the development of content and to put in place technical conditions in order to facilitate the presence and use of all world languages on the Internet;
- to ensure that more than half the world's inhabitants have access to ICTs within their reach.[6]

The Digital Solidarity Fund and the Task Force on Financial Mechanisms

Adequate financing represented the key obstacle between the benevolent intentions of the international community and the achievement of the goals mentioned above. The integration of new information and communication technologies into disadvantaged areas often requires the development of basic telecommunications infrastructure alongside advanced ICTs. This is expensive. Training of workers to set up and maintain these systems, and of users to operate them, is an often overlooked, but crucial and significant additional cost.

The digital divide was essentially agenda item one at WSIS and the issue of financing was a focus throughout. As the existing response to the digital divide was discussed and the politics around the issue were gradually fleshed out, it became clear that the WSIS would produce one of four outcomes on this topic:

- Developed states simply would declare current mechanisms adequate at the beginning of the WSIS and maintain the *status quo* by refusing to commit themselves to new expense. Such a decision would be seen as an explicit refusal on the part of developed states to support ICTD and would be difficult to justify to the international community, in particular given the involvement of civil society at the summit.
- Good-faith evaluation and auditing of existing international financial mechanisms would be undertaken in the aim of improving their reach, impact and efficiency.
- The development of new financial transfer mechanisms to replace existing ones would be pushed on the WSIS agenda, testing the resolve of the Global North to remain noncommittal.
- The evaluation and revision of existing mechanisms would take place *as well as* the creation of additional new financial transfer mechanisms.

Though it became apparent early in the Geneva phase that any negotiations concerning new financial transfers that implied some form of increase to

---

6   WSIS Executive Secretariat, *Geneva Plan of Action* (WSIS-03/GENEVA/DOC/5-E). (December 12, 2003). http://www.itu.int/dms_pub/itu-s/md/03/wsis/doc/S03-WSIS-DOC-0005!!MSW-E.doc

the financial contributions of developed countries would be met with serious political resistance from certain government delegations, backed largely by developing countries (with significant support for sectors of CS), the WSIS discussions were pushed in the direction of the latter of these options. The Digital Solidarity Fund (DSF), a decidedly new financial mechanism and one of the most interesting initiatives proposed at the WSIS, was met with a lukewarm reception from delegations from developed countries during the Geneva phase.

The Establishment of the Digital Solidarity Fund

The goal of the DSF was to engage as many international partners as possible (particularly governments of the developed North) and have them demand that various categories of networking, IT and communications firms contribute 1% of the total value of certain government contracts to a fund that would be earmarked for subsidizing connectivity in marginalized communities and regions. The "1% digital solidarity principle...is neither a tax nor a donation, but an investment in the markets of tomorrow" and must, it was argued, "become a universal principle for an equitable information society." Supporters based this claim on the rationales that:

> It complements traditional development funding by offering a stable source of revenue, specifically intended to reduce the digital divide.
>
> Based on the voluntary commitment of public institutions or private companies, it offers everyone an opportunity to take concrete action to build a more equitable information society.
>
> Deducted from the supplier's profit margin, it entails no direct cost to the institution or company that applies it.
>
> Clearly specified in the call for bids, its application is straightforward and unambiguous. Therefore, it respects the rules of free competition.[7]

The goals of the DSF were to ensure affordable and equal access to all ICTs as well as their content, to contribute to the social, political, economic, and cultural development of individuals and communities, as well as to reduce cultural and economic inequalities between individuals and social groups by redistributing the financial resources necessary for reducing the digital divide.

Government reactions to the proposal ranged from tepid to hostile and the Declaration of Principles adopted in Geneva stopped at acknowledging:

---

7   Digital Solidarity Fund website. http://www.dsf-fsn.org/cms/content/view/39/73/ lang,en/

the will expressed by some to create an international voluntary "Digital Solidarity Fund," and by others to undertake studies concerning existing mechanisms and the efficiency and feasibility of such a Fund.

The Plan of Action, also adopted during the first phase, was equally unenthusiastic about the issue, expressing a prudent attitude towards it:

> f) While all existing financial mechanisms should be fully exploited, a thorough review of their adequacy in meeting the challenges of ICT for development should be completed by the end of December 2004. This review shall be conducted by a Task Force under the auspices of the Secretary-General of the United Nations and submitted for consideration to the second phase of this summit. Based on the conclusion of the review, improvements and innovations of financing mechanisms will be considered including the effectiveness, the feasibility and the creation of a voluntary Digital Solidarity Fund, as mentioned in the Declaration of Principles.

Once Senegal recognized the lack of enthusiasm of certain heads of state, it turned for support to the cities and local authorities convening in Lyon at the World Summit of Cities and Local Authorities on the Information Society. The participants in this summit, which was held concurrently with Phase I of the WSIS in December 2003, supported the initiative.

As was the case in Geneva when it was proposed, governments refused to commit themselves to officially support the Digital Solidarity Fund during the Tunis phase. However, after the initiative was legally established in 2004 based on the support of the cities of Geneva and Lyon, PrepCom II of Phase II formally recognized the Fund in February 2005 and the Tunis Agenda for the Information Society specifies that:

> **28. We welcome the Digital Solidarity Fund (DSF)** established in Geneva as an innovative financial mechanism of a voluntary nature open to interested stakeholders with the objective of transforming the digital divide into digital opportunities for the developing world by focusing mainly on specific and urgent needs at the local level and seeking new voluntary sources of "solidarity" finance. The DSF will complement existing mechanisms for funding the Information Society, which should continue to be fully utilized to fund the growth of new ICT infrastructure and services.[8]

The Digital Solidarity Fund, which encouraged states to make "voluntary" contributions, managed to sidestep the cool reactions of governments and gain support from different actors, especially private enterprises, local public institutions, regions, and administrative departments. According to Alain Clerc, executive secretary of the DSF (and former director of the Civil Society Divi-

---

8   WSIS Executive Secretariat, *Tunis Agenda for the Information Society* (WSIS-05/TUNIS/DOC/6 (rev. 1)). (November 18, 2005). http://www.itu.int/wsis/docs2/tunis/off/6rev1.doc

Division of the WSIS Executive Secretariat), the DSF marked a series of important innovations on the international scene despite the refusal of many states to endow it with stable funding and a concrete mandate.

> It is the first time that an initiative brought forth by local authorities has been subsequently endorsed by Governments. More than just a rare and successful ingress of the local into the global, it is a significant turnaround of the situation, with a commitment of local communities to fully participate in international efforts for development.
>
> It is the first time that a new financing mechanism has been created to respond to the challenges of the Millennium Declaration (raising an additional 60 billion dollars) and proven to be effective. The Tobin Tax continues to miss the mark, President Lula's Fund, as attractive as it may seem, is having a hard time materializing, and other solidarity funds, though voted unanimously by the United Nations General Assembly, have yet to prove operational.[9]

Despite the Fund's voluntary nature, its recognition by government delegations in Tunis was indeed an accomplishment. Yet, overall, the inability of the Digital Solidarity Fund's promoters to rally the support of the countries of the North was a crushing blow. Furthermore, the DSF's potential for impact was seriously limited by its unsystematic and non-binding payment structure. The fund was reported to be in crisis by April 2009 for lack of funds, and its future is very much in doubt.[10]

Task Force on Financial Mechanisms: Activities and Final Report

Debates on the issue of financing raised fears of political failure at the WSIS and were postponed to the second phase when persisting disagreements made it obvious that a political consensus could not be reached during the first phase. Following the end of the first phase, a Task Force on Financial Mechanisms (TFFM) was created to draft a report on the issue. Political negotiations resumed only after the submission of this report. The work of the TFFM framed political negotiations by identifying the principal financial mechanisms in place and evaluating both their efficiency and relevance. Section 27(f) of the Geneva Plan of Action mandated the Secretary-General of the United Nations to head the establishment of the TFFM.

---

9   Alain Clerc, "Innovative Financial Mechanisms, Digital Solidarity and the 'Geneva Principle.'" In Daniel Stauffacher & Wolfgang Kleinwächter (eds.), *The World Summit on the Information Society: Moving from the Past Into the Future*. New York: United Nations Information and Communication Technology Task Force, 2005 (at 181).

10 Daniel Pimienta, *Digital Solidarity Fund in Risk of Disappearing: Action Required*. [WSIS CS-Plenary]. (April 13, 2009).

The United Nations Development Programme (UNDP), in collaboration with the World Bank and the UN Department of Economic and Social Affairs, was given the task of organizing the work of the TFFM in preparation for the report submission deadline at the end of 2004. Although relevant research was gathered prior to the initiative's official launch, the TFFM's working period officially lasted only two months. Difficult time constraints, compounded with a limited mandate and a top-down, opaque, and exclusive organizational structure, discredited the process and weakened the conclusions outlined in the report presented in December 2004.

Though it was officially launched in the spirit of openness and consultation, the TFFM came to be strongly criticized by civil society organizations, which were virtually excluded from the group. Civil society was only given a minimal presence in the task force, the report drafting work was undertaken even before the official launch of the TFFM, and the TFFM member appointment procedure was top-down and not very transparent. Contrary to the WGIG, for example, the TFFM generally turned out to be a disappointing multi-stakeholder experience marked by a lack of inclusion, openness, and transparency

Although electronic consultation processes were put into place and informal meetings were organized, the work of the TFFM remained opaque to say the least, given the pressure resulting from an extremely tight schedule and a limited mandate. After disappointing results from a consultation session, the Association for Progressive Communications (APC) addressed an open letter to the president of the TFFM in which it expressed its concerns. The APC notably deplored the severe time constraints imposed on the TFFM, the lack of transparency in the organization of the group's activities, the lack of multi-stakeholder participation mechanisms within the TFFM, the liberal bias of the report published by the group, and the group's inability to approach new financing mechanisms.[11] Numerous civil society associations supported this letter.

The concerns raised by the APC demonstrate the extent to which the issues of participation and content were linked at the WSIS. Having been largely marginalized at the TFFM, civil society did not get a proper chance to voice its opinions within the task force, and thereby lost access to one of the most significant political negotiation-orienting sites at the summit.

---

11 Anriette Esterhuysen, *An Open Letter on APC's Concerns about the Task Force on Financial Mechanisms, Process, Draft Findings and Conclusions.* (December 7, 2004). http://www.apc.org/apps/img_upload/5ba65079e0c45cd29dfdb3e618dda731/TFFM_Open_letter_from_APC.pdf

However, the TFFM did describe several important elements that will be presented here.

First of all, the TFFM recognized the contributions of new information and communication technologies to the international development process. According to the TFFM:

> ICTs are rapidly emerging as a vital factor in economic and social development to facilitate innovative and scalable solutions for achieving major development objectives. The potential for ICTs to have a decisive impact on achieving fundamental development goals, including those articulated in the Millennium Declaration, is increasingly recognized. Information and ICT-enabled services can serve to increase economic opportunities for the poor and disadvantaged, creating prospects for new jobs and small businesses along with increased knowledge to be applied in enhancing traditional livelihoods. Women stand to gain by being empowered through access to communication and learning networks. Health care systems can be vastly more effective. Learning can be enhanced and access to education made more equitable. Governments can provide more efficient and transparent services and respond to public needs more directly. The media and citizens are also able to empower themselves and become key players in local and national governance issues.[12]

The TFFM also recognized that NICTs applied in the service of development would require considerable financial resources and would not always guarantee meaningful returns on investments.

The report's conclusions were evocative and upheld the role of the private sector as central to the elimination of the digital divide. It was argued that bridging the digital divide required maximizing private sector investment in connectivity in the developing world through the creation of an enabling environment.

Such neo-liberal economic discourses centering on private sector leadership and the development of a competitive, transparent, safe, and predictable environment dominated the TFFM report and were seen to define its agenda.[13] Seen from this perspective, the role of developing countries is centered on creation of a favourable climate for private sector investment and the report goes on to suggest the implementation of policies aimed at granting universal access to these technologies as a vehicle for greater integration of economically disadvantaged countries into competitive global economies.

Governments are also encouraged to adopt "e-strategies" that cater to the needs of their populations, evaluate existing gaps, propose development and

---

12 For a complete list of TFFM members and other participants, see the final report (at page 95). Task Force on Financial Mechanisms for ICT for Development, *Financing ICTD–A Review of Trends and an Analysis of Gaps and Promising Practices*. (December 22, 2004). p.2. http://www.itu.int/**wsis**/tffm/final-report.pdf www.itu.int/**wsis**/tffm/final-report.pdf

13 Ibid.

implementation programs, and develop the social and economic policies required for establishing a favorable environment for developing new information and communication technologies. But the TFFM's report also confirmed the multi-stakeholder principle and reiterated the importance of taking inclusion-oriented action in order to integrate the interests and perspectives of the various stakeholders concerned with the implementation of development and access policies for NICTs.

The report conclusions focused on: ensuring the full use of current financing mechanisms, ensuring their adequacy, and recommending improvements and innovations. The TFFM specifically called for the development of a multi-stakeholder approach, and better coordination of existing financing mechanisms, initiatives, resources, and energy. It also recommended greater consideration of certain issues that are clearly associated with eliminating the digital divide but that, at present, are largely neglected by international institutions and private sector firms, a list that includes issues such as: sensitizing the population to NICTs through education and training and the question of integrating infrastructures into markets that are not attractive to private sector investment.

The TFFM's work was limited to evaluating existing financial mechanisms as mandated by the Geneva Plan of Action. However, it was also tasked with evaluating the relevance and feasibility of developing a Digital Solidarity Fund. The TFFM refused to do this, maintaining that it would be impossible to evaluate the DSF's effectiveness on reducing the digital divide because the Fund was not operational at the time that the final report was submitted. The TFFM's refusal to evaluate the DSF was strongly criticized by civil society, which supported the DSF and understood the stakes of its official recognition.

Interestingly, despite the TFFM's emphasis on market forces as the means for financing development infrastructure, equipment, and workforce training in developing countries, the Task Force also emphasized the public nature of the knowledge delivered by NICTs. This approach, similar to the approach of some civil society organizations, conceptualized knowledge and information as public goods, suggesting that access to, or use of them, by one party does not diminish any other party's ability to benefit from them equally:

> The characteristics of knowledge as a *public good*, and the role that ICT networks play in facilitating production and access to it, has strengthened support in various quarters for making access to ICT networks widely available. This is also because, as is the case with other *network* technologies, as more regions and actors are integrated into the network, benefits from their use for commercial and non-commercial uses grows, suggesting that everyone stands to benefit from investments in and expansion in access to ICT and ICT-enabled services. The role of ICT networks as a public good is also emphasized with regard to the range of services it can help to deliver. Both indi-

rectly through the public goods lens and directly, there has been growing interest and research into the means, both direct and indirect, by which ICT can help attain key development objectives and contribute to the achievement of the MDGs.[14]

Civil society gathered at the second PrepCom of the Tunis phase demonstrated a profound dissatisfaction with the TFFM's work and conclusions. Various civil society bodies worked together to draft a noticeably different vision of information society financing for development. This vision was presented to government delegations:

> We believe that the following principles need to form the basis of any discussion on financing the information society.
>
> **1. Information and communications and networks are a global public good**
> We believe that financing the information society should be based on the principle that information and communication is a global public good. This is particularly true for the extension of network infrastructure in developing countries and to all excluded populations everywhere. The value of global information networks increase as more national networks and individual users are added.
>
> **2. Centrality of the role of public finance**
> We recognise that private investment is an unique opportunity on a scale that in many ways is unique to ICD (Information and Communications for Development). However, private investment cannot displace the central role of public finance in a core development area like ICD.
>
> While encouraging the role of private investment in meeting the goals of ICD, the limits of the market in reaching these goals must be recognised.
>
> Public resources need to be mobilised at local, national and international levels.
>
> **3. The role of community driven and owned initiatives**
> The potential of community driven and owned ICD initiatives to contribute to sustainable development and social empowerment, especially women's empowerment, should be explored and integrated into financing strategies.[15]

Addressing the government plenary in the name of the European Civil Society Caucus on February 17, 2005, Steve Buckley summarized a more CS-focused perspective on the issues associated with the financing of the information society:

---

14 Ibid (at 15).
15 Anita Gurumurthy, *Statement Read by Anita Gurumurthy on Financing the Information by the Association for Progressive Communications, Bread for All, CRIS, Instituto del Tercer Mundo (ITeM), IT for Change and the Gender Caucus.* (February 15, 2005). http://www.itu.int/wsis/docs2/pc2/subcommittee/IT4ChangeDAWN.pdf

Governments of the north and south have a responsibility to engage with civil society to ensure effective implementation but this requires addressing our different perspective on the Plan of Action. The Political Chapeau and Implementation Plan of the Tunis phase is an opportunity to retake this discussion and, in particular, to ensure that the Action Plan and its implementation is oriented towards the implementation of internationally agreed human rights standards and internationally agreed sustainable development goals.

This includes ensuring that investment is oriented towards a vibrant civil society capable of holding governments to account, defending human rights and empowering people and communities. This includes ensuring that commitments to 0.7 per cent development assistance are met. This includes ensuring that aid is not confused with trade. This includes ensuring that investment is oriented towards community-driven solutions. This includes substantive engagement in discussion on new and innovative financing mechanisms. This includes support for initiatives from the south such as the Digital Solidarity Fund. This includes addressing not only the digital divide, but also the communications divide including support for independent and community media and other civil society communications initiatives and appropriate technology solutions.[16]

The Question of Financing at WSIS: The Path Chosen by the International Community

The Tunis Agenda reaffirmed certain fundamental principles developed by the TFFM.

During Phase I of the WSIS, private investment and financing were positioned as the drivers of ICTD funding. As such, creating an investment-friendly climate conducive to private sector activity was seen as an important step towards eradicating the digital divide. This included liberalizing trade, achieving political stability, adopting policies that encouraged competitiveness and deregulation, and establishing public-private partnerships. Public financing was barely addressed by the Geneva Declaration of Principles, which proclaims nonetheless that:

> Policies that create a favourable climate for stability, predictability and fair competition at all levels should be developed and implemented in a manner that not only attracts more private investment for ICT infrastructure development but also enables universal service obligations to be met in areas where traditional market conditions fail to work. In disadvantaged areas, the establishment of ICT public access points in places such as post offices, schools, libraries and archives, can provide effective means

---

16   Steve Buckley, *WSIS PrepCom II Intervention by the Civil Society European Regional Caucus on the Draft Political Chapeau and Implementation Plan* (specifically addressing Chapter 2 on Financing). (February 2005). http://thepublicvoice.org/news/2005_ prepcom _int_ec.html

for ensuring universal access to the infrastructure and services of the Information Society.[17]

A loose coalition of developing country governments mobilized around the ICTD funding issue late in the Geneva phase and continued to push the agenda over the course of Phase II.

The Tunis Agenda largely pointed to private sector investment as the engine for economic development but, without seeking to question the decisive role of the private sector in financing and investing in ICTs in the developing world, a slightly more balanced approach that would underline the responsibilities of public authorities and the international community was sought during the Tunis phase and is reflected, to an extent, in the Tunis documents. The Tunis Agenda for the Information Society (at para 18), for example, recognizes that:

> market forces alone cannot guarantee the full participation of developing countries in the global market for ICT-enabled services. Therefore, **we encourage** the strengthening of international cooperation and solidarity aimed at enabling all countries.

In addition, the Tunis Agenda (at para 23) acknowledged that a number of important areas had not been adequately financed to date, including:

a) ICT capacity-building programmes, materials, tools, educational funding and specialized training initiatives, especially for regulators and other public-sector employees and organizations.
b) Communications access and connectivity for ICT services and applications in remote rural areas, Small Island Developing States, Landlocked Developing Countries and other locations presenting unique technological and market challenges.
c) Regional backbone infrastructure, regional networks, Network Access Points and related regional projects, to link networks across borders and in economically disadvantaged regions which may require coordinated policies including legal, regulatory and financial frameworks, and seed financing, and would benefit from sharing experiences and best practices.
d) Broadband capacity to facilitate the delivery of a broader range of services and applications, promote investment and provide Internet access at affordable prices to both existing and new users.
e) Coordinated assistance, as appropriate, for countries referred to in paragraph 16 of the Geneva Declaration of Principles, particularly Least Developed Countries and Small Island Developing States, in order to improve effectiveness and to lower transaction costs associated with the delivery of international donor support.

---

17 WSIS Executive Secretariat, *Geneva Declaration of Principles* (WSIS-03/GENEVA/DOC/0004) (December 12, 2003). http://www.itu.int/dms_pub/itu-s/ md/03/wsis/doc/S03-WSIS-DOC-0004!!MSW-E.doc

f) ICT applications and content aimed at the integration of ICTs into the implementation of poverty eradication strategies and in sector programmes, particularly in health, education, agriculture and the environment.[18]

The Tunis Agenda also recognized the need for countries experiencing structural difficulties related to their development and integration of NICTs to implement strategies that are appropriate for their particular situations, the urgent need for better allocation of resources and better coordination of actors and initiatives for reducing the digital divide, and the developed countries' responsibility to respect their international development commitments.

Elsewhere, the importance of technological transfers in eradicating the digital divide is cited four times in the Tunis Agenda (paragraphs 8, 49, 54, and 89). The transfers are a fundamental step towards including marginalized or excluded populations into large international communication networks. Technological transfers can take different shapes depending on circumstances and needs. First, according to a traditional model of cooperation, various technological tools that are rarely used or not used at all in the North can be transferred to the developing countries of the South. For instance, the increasingly high turnover of computers, cellular phones, radio transmitters, and recording equipment in the North could be transferred to the South at relatively low costs that would yield benefits for both parties. This would increase the life cycle of these tools and work in the interests of a large number of individuals and communities. The negative aspects of such transfers are that they would precipitate electronic pollution towards the countries of the South, many of which lack the necessary resources to properly dispose of the waste materials generated by such equipment. This remains a serious issue.

Technological transfers could also improve basic infrastructure supporting communication networks, especially with regard to the establishment of broadband networks using different satellite or cable technologies. Last but not least, transfers could also include software applications that serve as computerized tools with concrete applications for management, education, training, and communication. Free and open software programs would play an important role in this case. Such initiatives can be criticized for creating path dependencies but are one element of the Tunis phase outcomes that does express some political will to materially contribute to ICTD.

The governments that signed the Tunis Agenda did not meaningfully commit themselves to increasing their international contributions towards

---

18 WSIS Executive Secretariat, *Tunis Agenda for the Information Society* (WSIS-05/TUNIS/DOC/6 (rev. 1)). (November 18, 2005). http://www.itu.int/wsis/docs2/tunis/off/6rev1.doc

integrating developing countries into global communication networks. Despite being a major event that managed to bring together heads of state, the private sector, international organizations, and civil society to discuss the issue of digital inclusion for the first time, the WSIS did not succeed in convincing governments to make additional significant efforts in decreasing the profound inequalities that characterize the information society. The same summit actors who maintained that "we are indeed in the midst of a revolution, perhaps the greatest that humanity has ever experienced"[19] paradoxically refused to adopt initiatives that matched the complexity of the changes and challenges they claimed to be addressing.

Due to this refusal, it is the reorganization of already available resources and not an influx of new funds on which adequate information society financing, international aid and development, and integration of marginalized populations into global communication networks will depend going forward. This reality makes it even more important to monitor progress and hold governments accountable; it makes it more complicated as well.

Large international meetings are frequently criticized for producing texts that are diluted by the need for consensus and then further marginalized by a lack of commitment to see them applied. The WSIS sought to build a consensus on intersecting views of global governance by developing follow-up and implementation mechanisms that would transform declarations of intent into measurable and quantifiable initiatives. In this way, the challenges surrounding information society financing have become intrinsically linked with the follow-up and implementation measures adopted at the WSIS. It is here that it will be clear whether or not the political will exists to bridge the gap between intent and action, and principle and initiative. In a way, the extent to which the goals of these financing mechanisms can be achieved depends entirely on the adequacy of the implementation and follow-up mechanisms adopted at the summit.

Assessing Financial Mechanisms: Civil Society's Positions

Throughout the second phase of the WSIS, civil society was particularly critical of the consultation procedures set up at the summit for discussing issues of financing. Not surprisingly, CS was similarly displeased by the results produced within these processes. Civil society frequently objected to the government delegations' lack of vision as well as their refusal to commit themselves

---

19 Formerly published on the official WSIS website, this statement has now been removed.

to new financing mechanisms, and condemned the neo-liberal assumptions underlining certain positions defended and adopted at the summit.

CS celebrated the launch of the DSF and took credit for introducing language on the "importance of public policy in mobilizing resources for financing" into the Tunis Agenda (at paras 21 and 35) as a measure of balancing the discourse of market-based solutions that dominates the Agenda elsewhere. But, CS was more critical of what was not said at Tunis than what was, using the CS Phase II declaration to suggest that:

> Investments in ICTD—in infrastructure, capacity building, appropriate software and hardware and in developing applications and services—underpin all other processes of development innovation, learning and sharing, and should be seen in this light. Though development resources are admittedly scarce and have to be allocated with care and discretion, ICTD financing should not be viewed as directly in competition with the financing of other developmental sectors. Financing ICTD should be considered a priority at both national and international levels, with specific approaches to each country according to its level of development and with a long-term perspective adapted to a global vision of development and sharing within the global community.
>
> Financing ICTD requires social and institutional innovation, with adequate mechanisms for transparency, evaluation, and follow-up. Financial resources need to be mobilized at all levels—local, national and international, including through the realization of ODA commitments agreed to in the Monterrey Consensus and including assistance to programs and activities whose short-term sustainability cannot be immediately demonstrated because of the low level of resources available as their starting point.

And to make the case that:

> Internet access, for everybody and everywhere, especially among disadvantaged populations and in rural areas, must be considered as a global public good.[20]

Largely excluded from the work of the TFFM, civil society proposed an alternative model to the neo-liberal option upheld by the Task Force and later adopted by government delegations. This alternative, characterized by a particular "emerging paradigm" on information society financing for development, was radically different from the vision adopted by government delegations. Centered on the notion that ICTs are "global public goods," the model planned for governments to finance network extensions with the new

---

20 This document is included in the appendix of the present volume. Full citation: WSIS Civil Society Plenary, *Much More Could Have Been Achieved: Civil Society Statement on the World Summit on the Information Society.* (December 18, 2005 Revision 1–December 23, 2005). p.4–5. http://www.worldSummit2003.de/download_en/WSIS-CS-Summit-statement-rev1-23-12-2005-en.pdf

revenue that these public goods could generate. Governments played an important role in this perspective, as they would have to ensure their public interest principle by creating an environment that would satisfy everybody's communication needs, or in other words, providing free and extensive access to knowledge and information. According to one prominent CS activist:

> These principles and practices are proposed not as a replacement for market driven ICT development, which will have a continuing and dominant relevance. But they do embody a deep criticism of the market driven approach, one that goes beyond an admission of temporary 'market failure' at the margins. At the same time, they offer a flexible, realistic, development-driven dynamic, and claim it can be generated alongside the market situation.[21]

The perspective on financing developed by civil society at the WSIS was ultimately intended to counterbalance a consensus made by government delegations at the international level. This consensus, far from being based on international aid, or on the mobilization of governmental resources in the service of socioeconomic development goals, was more in line with a rhetoric that linked private investment closely with economic development.

The differences between the positions defended by civil society and the neo-liberal vision preferred by governments ran deeper than merely divergent views on methods that ought to be used to evaluate financing mechanisms and the extent to which they catered to the needs of populations that were marginalized from global communication networks. Although these opposing views were partly based on different ethical conceptions of the notion of international development, their foundations lay mainly in the notion of *global social solidarity*. Rather than viewing the inclusion of marginalized actors into computerized communication networks as a goal that was to be achieved, the principle of solidarity saw their inclusion as the foundation of a new social contract for a global information and knowledge society.

This view critically recognizes that the intensification and quantitative enlargement of communication techniques and the increasing speed at which they are being developed has led to forecasts of a qualitative change from a social order based on the industrial model to one based on an informational model. This supposed transformation from an industrial to an information society was the very essence of the political project encapsulated by the WSIS. In the West, industrial society has historically been legitimized by the development of social programs that, at least officially, aim to redistribute the re-

---

21 Seán Ó Siochrú, *Mapping Research Activities and Interactions Against Global Communication Dynamics: The Case of the WSIS and Financing Mechanisms*. (n.d.). http://www.is-watch.net/node/1024

sources necessary for the social and economic inclusion of disadvantaged populations. According to civil society actors, the refusal of government delegations to subscribe to a vision of information society financing that perceives access to knowledge networks as the information society's main social and moral legitimizing force discredited the social and economic project that had been put forward at the WSIS.

Thus, the social solidarity principle put forward by civil society at the WSIS framed the issue of information society financing not as a mere tool for international development, but as the very basis of the social legitimacy of a globalized information society. Access to knowledge, information, and culture is the basis for social, economic, and political integration in this approach.

• CHAPTER FIVE •

A Geopolitics of Networks: Internet Governance at WSIS

Neither Internet regulation in the broad sense, nor the more narrowly defined issue of the governance of the Internet domain name system (DNS), figured prominently in the plans or initial agenda of the WSIS. This chapter reviews how, through the chaotic process through which delegations and summit organizers sorted out a program for a policy object as abstract as the 'information society', a series of competing problems and claims coalesced, in various and changing combinations, to push Internet regulation and governance up the WSIS agenda. It will examine the fluid and inconsistent manner in which the issue of Internet governance was framed over the course of the first phase of the WSIS and the intractable debate that followed over the extent to which broader public policy issues apply to Internet governance. Through WSIS, Internet governance came to be seen as encompassing a much wider range of issues and concerns than merely domain name management.

In defining the term going forward to the second phase of the WSIS, lines were drawn about what issues were considered to fall under the heading of global Internet governance. However, the mere recognition of the public policy implications of Internet governance is not the end of the story. A focused—but equally charged and controversial—debate occurred as Internet governance came to dominate the agenda of the second phase of the WSIS. This debate was marked by a sensationalized split between the US and the EU on the issue of governmental oversight of the Internet Corporation for Assigned Names and Numbers (ICANN), and the dynamics of the Tunis deal that brokered compromise through the creation of the Internet Governance Forum—a new multi-stakeholder global governance organization—as well as an ill-defined process of "enhanced cooperation."

Internet Governance at WSIS Phase I: [International/Intergovernmental]

ICANN is a private organization that coordinates the naming and numbering registry that allows the global Internet to function. Operating through a series of agreements with the Department of Commerce of the United States Government, ICANN was the controversial outcome of a decade-long discussion over how the Internet domain name system should be governed.[1]

ICANN is an essentially unique experiment in global governance.[2] It is a not-for-profit corporation based in Marina Del Ray, California, incorporated under the statutes of the State of California. ICANN coordinates the DNS system, the root server system, the allocation of IP addresses and policy development related to these functions.[3] It does so through what it describes as a bottom-up, multi-stakeholder policy development process that is typically framed as an extension of the standards-making processes the Internet technical community has developed through structures such as the Internet Architecture Board (IAB) and Internet Engineering Task Force (IETF). ICANN has been a lightning rod for criticism in its short history. For example, it is seen as prone to being captured by the agendas of technical elites, the US government and multi-national communication firms in particular.[4] More generally, it is seen—simply put—as not sufficiently democratic or accountable.[5] ICANN has always insisted that its role is neutral, technical management. As Esther Dyson, the first president of ICANN, once quipped: "ICANN governs the plumbing not the people."[6]

From the first WSIS PrepCom there were indications that discord over the role of governments in decision-making related to the management of the

---

1 For more detailed background see Milton Mueller, *Ruling the Root: Internet Governance and the Taming of Cyberspace*. Cambridge, Mass: MIT Press, 2002; Daniel Paré, *Internet Governance in Transition: Who Is the Master of This Domain?* Lanham, MD: Rowman & Littlefield, 2003; Jack Goldsmith and Tim Wu, *Who Controls the Internet? Illusions of a Borderless World*. New York: Oxford University Press, 2006.

2 See Sean Ó Siochrú and Bruce Girard, *Global Media Governance: A Beginner's Guide*. Boulder & London: Rowman & Littlefield, 2002.

3 See Jeanette Hoffmann. "ICANN." in *Global Information Society Watch 2007*. Uruguay: Association for Progressive Communications and Third World Institute, 2007.

4 Milton Mueller, *Ruling the Root: Internet Governance and the Taming of Cyberspace*. Cambridge, Mass: MIT Press, 2002.

5 Jack Goldsmith and Tim Wu. *Who Controls the Internet? Illusions of a Borderless World*. New York: Oxford University Press, 2006 (at 70).

6 Quoted in Milton Mueller, *Ruling the Root: Internet Governance and the Taming of Cyberspace*. Cambridge, Mass: MIT Press, 2002 (at 8-9).

Internet was going to creep its way onto the WSIS agenda. Brazil's intervention at PrepCom I of Phase I complained that

> democratic and representative Governments should not be replaced by arbitrary groupings of private business and non-governmental institutions in decisions regarding the economic space brewing within powerful digital networks, such as the Internet. Organizing this new environment to the satisfaction of all, and ensuring the beneficial participation of developing countries and their societies is central to our work.[7]

The EU's contribution was both more specific and instrumental, calling for the WSIS "to indicate a set of common principles underlying future actions and initiatives" related to "electronic communications regulatory frameworks" as well as "legal aspects of e-commerce and Internet governance."[8]

Slotted between PrepCom I and PrepCom II, the International Telecommunication Union (ITU) held its regularly scheduled Plenipotentiary Conference in Marrakech in October of 2002. According to Kleinwächter, "a bitter controversy about private sector leadership and the future role of ITU in Internet governance took place."[9] A series of resolutions were passed underlining the need to reinforce the sovereignty of governments in domain name related matters and directing the ITU secretary general to encourage all ITU member states to participate in discussions over international management of domain names and numbers. These resolutions also encouraged the secretary general of the ITU to himself take a significant role in such debates and initiatives.[10]

From there, the issue of Internet governance percolated onto the WSIS agenda through a series of declarations made by regional WSIS preparatory meetings that were also held between PrepComs I and II. The declaration of the European WSIS Ministerial Meeting in Bucharest (Nov. 2002) mentions in passing that "management of domain names" is one among many issues

---

7   WSIS Executive Secretariat, *Geneva Phase, PrepCom-1: Statement from Brazil.* (July 1–5, 2002). http://www.itu.int/wsis/docs/pc1/statements_general/brazil.doc
8   European Union, *The UN Summit on the Information Society: The Preparatory Process. Reflections of the European Union* (WSIS/PC-1/C/0003). (June 19, 2002). http://www.itu.int/dms_pub/itu-s/md/02/wsispc1/c/S02-WSISPC1-C-0003!!MSW-E.doc
9   Wolfgang Kleinwächter, "Beyond ICANN vs. ITU: Will WSIS Open New Territory for Internet Governance?" in Don MacLean (ed.), *Internet Governance: A Grand Collaboration.* New York: UN ICT Task Force, 2004 (at 42).
10 International Telecommunication Union, *Resolution 133 of the 2002 Marrakech ITU Plenipotentiary Conference: Role of administrations of Member States in the management of internationalized (multilingual) domain names.* (2002). http://www.itu.int/aboutitu/basic-texts/resolutions/res133.html

that should "be addressed with the active participation of all stakeholders."[11] The declaration of the January 2003 Asian WSIS Ministerial Conference in Tokyo declares that

> the transition to the information society requires the creation of appropriate and transparent legal, regulatory and policy frameworks at the global, regional and national levels. These frameworks should give due regard to the rights and obligations of all stakeholders.

The subsequent list of policy issues requiring such reforms includes "management of Internet addresses and domain names."[12] Two weeks later, the Latin American Regional Ministerial Conference held in Bávaro called for "multi-lateral transparent and democratic Internet governance" as part of an effort to establish "appropriate national legislative frameworks that safeguard the public and general interest and intellectual property that foster electronic communications and transactions."[13] Finally, in February 2003, in Beirut, the West Asia Ministerial Conference for WSIS mentioned "multilingualism" and "national sovereignty" as two of the reasons that "the responsibility for root directories and domain names should rest with a suitable international organization."[14]

Thus, the controversies emerging around ICANN outside of the main WSIS plenary made the WSIS a logical venue for raising the issue of Internet governance. Yet, as discussed, the WSIS Phase I agenda was far from set and domain name system problems were competing for limited attention alongside much broader Internet, ICT and information society public policy issues. By PrepCom II of Phase I, ICANN issues were being raised at WSIS, but often interchangeably with, and indistinguishably from, broader concerns about the governance of communication and the public's interest in it. A Brazilian intervention to PrepCom II for instance argued that

> Internet has evolved into a global public good and its governance should constitute a core issue of the information society agenda. Developing countries should have full

---

11 WSIS Executive Secretariat, *Final Declaration of the Pan European Regional Conference* (WSIS/PC-2/DOC/5-E). (November 9, 2002). http://www.itu.int/wsis/documents/listing-all.asp?lang=en&c_event=rc|pe&c_type=all|

12 WSIS Executive Secretariat, *Final Declaration of the Asia-Pacific Regional Conference* (WSIS/PC-2/DOC/6-E). (January 22, 2003). http://www.itu.int/wsis/documents/listing-all.asp?lang=en&c_event=rc|as&c_type=all|

13 WSIS Executive Secretariat, *The Bávaro Declaration* (WSIS/PC-2/DOC/7-E). (February 5, 2003). http://www.itu.int/wsis/documents/listing-all.asp?lang=en&c_event=rc|l&c_type=all|

14 WSIS Executive Secretariat, *The Beirut Declaration* (WSIS/PC-2/DOC/8-E). (February 5, 2003). http://www.itu.int/dms_pub/itu-s/md/03/wsispc2/doc/S03-WSISPC2-DOC-0008!!MSW-E.doc

access to take part in all decision-making bodies and processes concerning the structure and functioning of the cyberspace, within which public, private and nongovernmental agents will increasingly conduct their social and economic activities.[15]

Supporters of ICANN largely chose to ignore calls that the issue of its reform be placed on the WSIS agenda, presumably hoping that the issue would simply go away. By PrepCom II, some veiled support for the existing institutional framework for Internet governance was evident in certain interventions. The private sector Coordinating Committee of Business Interlocutors (CCBI) contribution to PrepCom II offered a carrot in the direction of calls for more intergovernmental cooperation on Internet governance: "many cross-border issues have already been and are being coordinated by international fora." [16] But also a stick:

> the critical role of the private sector must be recognized more clearly and actively in the WSIS process. This has not been adequately done to date. [...] Given the right conditions, business will assume the risks necessary, and invest. [...] 'No investment, no information society'.[17]

Updated drafts of the Plan of Action and Declaration of Principles were circulated on March 21, 2003, and comments were solicited from all stakeholders in advance of the "intersessional" meeting planned for Paris in July. The March 2003 draft documents reformulated and reorganized the language on Internet governance slightly. The new formulation in the plan of action was particularly revealing. At paragraph 33, the proposed plan of action reads:

> **Internet governance:** A transparent and democratic governance of the Internet shall constitute the basis for the development of a global culture of cyber-security. An [international] [intergovernmental] organisation should ensure multilateral, democratic and transparent management of root servers, domain names and Internet Protocol (IP) address assignment.[18]

---

15 WSIS Executive Secretariat, *Geneva Phase: PrepCom-2: Contribution: Governments: Brazil.* (WSIS/PC-2/CONTR/57-E). (January 7, 2003). http://www.itu.int/dms_pub/itu-s/md/03/wsispc2/c/S03-WSISPC2-C-0057!!PDF-E.pdf

16 Coordinating Committee of Business Interlocutors (CCBI), *What Are the Contents and Themes that Business Supports for the Summit?* (WSIS/PC-2/CONTR/35-E). (December 10, 2002). http://www.itu.int/dms_pub/itu-s/md/03/wsispc2/c/S03-WSISPC2-C-0035!MSW-E.doc

17 Coordinating Committee of Business Interlocutors (CCBI), *Business View on Summit Outcomes* (WSIS/PC-2/CONTR/35-E). (December 14, 2002). http://www.itu.int/dms_pub/itu-s/md/03/wsispc2/c/S03-WSISPC2-C-0035!A1!MSW-E.doc

18 WSIS Executive Secretariat, *Draft Action Plan Based on Discussions in the Working Group of Sub-Committee 2* (WSIS/PCIP/DT/2-E). (March 21, 2003). http://www.itu.int/dms_pub/itu-s/md/03/wsispcip/td/030721/S03-WSISPCIP-030721-TD-GEN-0002!!MSW-E.doc

That the words "international" and "intergovernmental" are inserted into para 33 in square brackets is significant. In the referencing system used to mark-up negotiation documents in the UN system, square brackets are commonly used to indicate text that has been suggested by one or more delegations but that lacks unanimous approval.[19] By the intersessional meeting, in other words, it was clear that some WSIS delegations were campaigning for reform of the ICANN model or for its replacement by a more traditional, intergovernmental organization while others supported the existing model of an international organization that is not intergovernmental. The first phase of the debate over Internet governance would be a battle over which word was going to have its square brackets removed. What was not clear was whether this debate was only about the DNS system or what exactly Internet governance meant in the context of the WSIS. After lurking at the margins for the first months of the WSIS, by the end of PrepCom II, Internet governance had officially arrived on the WSIS agenda, whatever that meant.

In their comments on this draft of the Declaration of Principles, Cuba suggested inserting the word "intergovernmental" into the first sentence of paragraph 44 so that it would read: "Internet governance must be multilateral, intergovernmental, democratic and transparent."[20] Brazil was more explicit, suggesting that its previously cited comments about the Internet as a public good be amended to acknowledge that:

> The International Telecommunication Union, as a specialized agency of the United Nations System, shall play a leading role in the emergent information society and in the regulation of the global information and communications infrastructure.[21]

By this point it was becoming clear that, for Brazil, an intergovernmental organization was required to govern a global public good. What was less clear at the time, however, was whether the global public good was, in Brazil's view, the DNS system or communication on the Internet.

For example, in their comments on the Plan of Action, Brazilian-proposed amendments included:

---

19 For general scholarly analysis of UN document practices see Annelise Riles (ed.), *Documents: Artifacts of Modern Knowledge*. Ann Arbor: University of Michigan Press, 2006.

20 WSIS Executive Secretariat, *Reference Document —Compilation of Contributions to the Draft Declaration of Principles and Draft Plan of Action–Part I (Declaration of Principles), Section I Governments' contributions* (WSIS/PCIP/DT/3-E). (June 12, 2003). http://www.itu.int/dms_pub/itu-s/md/03/wsispcip/td/030721/S03-WSISPCIP-030721-TD-GEN-0003!P1!MSW-E.doc

21 Ibid.

the Internet is the base of the information society. The Internet must be considered a public international domain. Every country and every person have the right to be connected and to take full advantage of the benefits offered by the Internet.[22]

However, Brazil then went on to insist—in the very same paragraph—that the administration of the DNS must occur through an intergovernmental organization and involve developing countries.[23] Again, the result is that when Brazil frames 'Internet governance' "as a key issue of the information society,"[24] it is entirely unclear whether Internet refers to the DNS system or something much broader.

The responses from developed country governments, the Internet technical community and the private sector supporting the status quo were varied.

The comments of certain delegations appeared to be aimed at diffusing the tension by replacing references to specific organizations and institutional labels with more general terms. Australia, for example, simply called for "administrative and coordination activities related to the Internet [to] remain the responsibility of *an organization* with broad stakeholder input" (emphasis added). Canada insisted that "the coordination responsible for root servers, domain names and Internet Protocol (IP) address assignment should rest with a *suitable* organization" (emphasis added). Leveraging the expressive potential of the conventions for marking up UN negotiating documents to the hilt, Japan's contribution was simply:

"An [international] [intergovernmental] organization should ensure multilateral...."[25]

'Broad', 'suitable', 'international': rather than referencing a definable institutional forum, each are vague normative catch-alls that could be argued to be present in the ICANN system as it was configured at the time, or to be a goal that the ICANN system was capable of working toward without fundamental changes in its mandate.

Other comments were slightly more restrictive and prescriptive. The US emphasized the need for "public-private partnership" in DNS management in order to "preserve and enhance the necessary global interoperability and coordination of the Internet's unique identifier system while recognizing its technical limitations and requirements."[26] The intervention of the Internet Society (ISOC)—a coordinating body for the activities and interests of the Internet

---

[22] Ibid.
[23] Ibid.
[24] Ibid.
[25] Ibid.
[26] Ibid.

technical community[27]—reinforced the implication that the calls for reform seemed to be neglecting the extent to which the characteristics of Internet technologies effectively constrain the range of policy alternatives. ISOC professed to be

> very concerned by statements in the draft documents that imply the need for new, intergovernmental organizations to "manage" the Internet. In particular, proposals to replace ICANN and create a new mechanism for managing root servers, domain names and IP addresses is unnecessary, will lead to significant disruption, and is unlikely to succeed.[28]

Intergovernmental oversight of the Internet was, in other words, a total non-starter to a series of delegations.

The invocation, at this juncture, of discourses on the limited extent to which the technology of the Internet would tolerate such efforts is revealing. By asserting that the power dynamics of global Internet governance simply preclude the possibility of dramatic calls for reform, the US, the ISOC and their sympathizers underline how, in the absence of a more intrusive legal framework, the ability to control and define technology is power. A subsequent CCBI intervention threatened that "business cannot accept any reference to an intergovernmental organization engaging in such management."[29] In other words, technological power was being wielded alongside political

---

27 Under the leadership of a group of engineers credited as the "founding fathers" of the Internet, often like-minded technologists have colacesed around a series of semi-institutionalized professional organizations. Some of these organizations, such as the Internet Engineering Task Force (IETF), are dedicated to cooperative work and problem solving around emerging Internet technical issues, such as standard setting. Others, most notably the Internet Society (ISOC), reflect a general effort to organize this community and lobby for its vision of the Internet and interests in Internet governance. All of these organizations are properly considered to be Internet governance bodies, in particular given the extent to which the lack of formal governance structures over the Internet devolves political control to technical management functions and those who operate them. Any possible reform of the existing IG system could impact this arrangement. To speak of the Internet technical community as a political entity as we have done here is, thus, in a broad sense to refer to the interests of those participating in this array of informal, technical governance bodies, those represented by and supporting the ISOC in particular. See Milton Mueller, *Ruling the Root: Internet Governance and the Taming of Cyberspace*. Cambridge, Mass: MIT Press, 2002.

28 ISOC, *Comments of the Internet Society on the World Summit on the Information Society (WSIS) Draft Declaration of Principles and Action Plan* (Document WSIS/PC-3/89-E). (May 31, 2003). http://www.itu.int/dms_pub/itu-s/md/03/wsispc3/c/S03-WSISPC3-C-0089!!MSW-E.doc

29 CCBI, *Comments on Draft Declaration of Principles and Draft Action Plan* (WSIS/PC-3/CONTR/10-E). (May 5, 2003). http://www.itu.int/dms_pub/itu-s/md/03/wsispc3/c/S03-WSISPC3-C-0040!!MSW-E.doc

economic power by the chief beneficiaries of the status quo in the effort to use their capacity leverage to bully the debate over meaningful reform of Internet governance right off the WSIS agenda before it even got off the ground.

The WSIS entered the so-called intersessional meeting (July 15-18, 2003, in Paris) with a singular, if contradictory and controversial, set of draft paragraphs on Internet governance. WSIS delegates thus arrived at the intersessional meeting entirely aware of the degree of divergence in opinions on Internet governance. In recognition that the WSIS was further away from reaching consensus on the language on Internet governance than it was on many of the other issues being discussed, governments created an "Internet Governance Ad-Hoc Working Group" at the intersessional meeting. With the exception of its first meeting, this working group did not adopt the multistakeholder rules of participation in force in the wider WSIS activities. Meetings of the IG Ad-Hoc Working Group were largely restricted to government delegations, even if many sympathetic government delegations chose to openly report back to civil society and private sector would-be interlocutors.[30]

By the conclusion of the intersessional meeting, the draft Declaration of Principles proposed three possible formulations of the main text on Internet governance. Each agreed that the "the international management of the Internet should be democratic, multilateral and transparent." Opinions, however, diverged from there.

One proposal recognized that Internet governance contained technical as well as policy issues. While private sector leadership should continue at the technical level, governments, it was argued, ought to

> take a lead role, in partnership with all other stakeholders, in developing and coordinating policies of the public interests related to stability, security, competition, freedom of use, protection of individual rights and privacy, sovereignty, and equal access for all.

This proposal remained unclear about whether this should occur within a traditional intergovernmental organization (presumably the ITU) or an 'international' one such as ICANN. The second proposal focused explicitly on the DNS. It asserted the sovereign rights of countries over policy authority of their CCTLDs and called for multilingualism in Internet governance. Responsibility for management of the DNS should reside, it continued, with an intergovernmental organization. The third proposed model suggested that global Internet governance should "respect geographic diversity" and ensure the par-

---

30 See Wolfgang Kleinwächter, "Beyond ICANN vs. ITU: Will WSIS Open New Territory for Internet Governance?" in Don MacLean (ed.), *Internet Governance: A Grand Collaboration.* New York: UN ICT Task Force, 2004.

ticipation of those governments that are particularly *"interested"* in Internet governance (emphasis added).[31]

The "Extract from the Draft Plan of Action" was already managing expectations about the prospects for agreement prior to the conclusion of the Geneva phase of the summit, suggesting that the second phase of the WSIS should be devoted to reviewing the continuing international dialogue on the subject.[32] In other words, by convening an ad hoc working group on the subject, the WSIS probably did more to reinforce the differences between delegations on the issue of Internet governance than it did to resolve them, at least initially.

Negotiation of Internet governance at PrepCom III was, according to Swiss diplomat Markus Kummer, "both very polarized and, to a large extent, also very abstract. There were misunderstandings on both sides."[33]

The 'governments only' edict of the Ad-Hoc Working Group on Internet Governance meant that many of the world's leading practitioners of, and experts on, Internet governance were left milling around in the corridors outside the rooms in which the negotiations were taking place when they very easily could have been called upon to help fill in knowledge gaps and offer explanations when it became clear that misunderstandings were holding up progress. In what has become an oft-repeated parable in WSIS civil society circles, even ICANN CEO Paul Twomey had to leave the room so that governments could resume their debate.

By the end of the first week of PrepCom III, a new (September 19, 2003) draft had broken out the three competing versions of the Declaration of Principles' paragraph 44 into a series of new paragraphs. The divisive point remained the question of how issues that were now being discussed under the decidedly broader-than-ICANN label of "Internet issues of an international nature related to public policies" should be coordinated. There were a series of alternative linguistic formations proposed, but the sticking point remained the

---

31 WSIS Executive Secretariat, *Draft Declaration of Principles Building the Information Society: a Global Challenge in the New Millennium* (WSIS03/PCIP/DT/4(Rev.3)-E). (November 14, 2003). http://www.itu.int/dms_pub/itu-s/md/03/wsispc3/td/030915/S03-WSISPC3-03 0915-TD-GEN-0004!!MSW-E.doc

32 WSIS Executive Secretariat, *Extract from the Draft Plan of Action* (WSIS03/PCIP/DT/7-E). (July 18, 2003) (at 104 and 105). http://www.itu.int/dms_pub/itu-s/md/03/wsispcip/td/030721/S03-WSISPCIP-030721-TD-GEN-0007!!MSW-E.doc

33 Markus Kummer, "The Debate on Internet Governance: From Geneva to Tunis and Beyond." *Information Polity* 12 (1–2), 2007 (at 6).

role of governments and the imposition of an intergovernmental institution dedicated to Internet governance.[34]

Over the rest of the September sitting of PrepCom III, however, these paragraphs would change only slightly,[35] and they would not change at all over the course of the first resumed PrepCom III session held in November 2003.[36]

In a stroke of nomenclature that could only have been produced within the UN system, it was decided that there would be a "PrepCom III resumed II" to take place immediately before the first phase of the summit, on December 5 and 6, 2003. Determined not to let the Geneva phase of the summit fail on its watch, the Swiss delegation, led by secretary of state for WSIS Marc Furrer, effectively took ownership over this final stage of the negotiation. Responsibility for pushing through a compromise on Internet governance was assigned to Markus Kummer.

Faced with five different proposals for the conclusion of the paragraph on "Internet issues of an international nature related to public policies" that he saw as "mutually exclusive" and confronted with delegations that "were firmly entrenched in positions that were diametrically opposed," Kummer concluded that "the only way out was to establish a process to deal with these issues." The last ditch efforts of PrepCom III resumed II focused less on bridging the gap between the different perspectives and instead "focused on the modalities of the process [delegates] hoped to initiate" to continue the discussion going forward.[37] A key domino of compromise fell when the secretary general of the ITU was replaced as the presumptive convener of the proposed follow-up study group by the Secretary-General of the UN itself. The initial proposal to appoint the head of the ITU as the chair of a group that was to study an issue being defined by a debate over the role of ICANN vs. the role of the ITU was politically fraught from the start. In Kummer's words, the formula of including some form of United Nations involvement without favoring calls for greater ITU leadership provided "the flexibility required to be inclusive" to both the intergovernmental (ITU) and private sector (ICANN) factions.

---

34 WSIS Executive Secretariat, *Draft Declaration of Principles* (WSIS/PC-3/DT/1). (September 19, 2003) (at para 44). http://www.itu.int/dms_pub/itu-s/md/03/wsispc3/td/030915/S03-WSISPC3- 030915-TD-GEN-0001!!MSW-E.doc

35 WSIS Executive Secretariat, *Draft Declaration of Principles* (WSIS/PC-3/DT/1(Rev.2B)-E). (September 26, 2003). http://www.itu.int/dms_pub/itu-s/md/03/wsispc3/td/030915/S03- WSISPC3-030915-TD-GEN-0001!R2B!MSW-E.doc

36 WSIS Executive Secretariat, *Draft Declaration of Principles Building the Information Society: A Global Challenge in the New Millennium* (WSIS/PC-3/DT/6-E (Rev.1)). (November 14, 2003). http:// www.itu.int/wsis/documents/listing-all-pc.asp?lang=en&c_event=pc|3

37 All quotes from Markus Kummer. "The Debate on Internet Governance: From Geneva to Tunis and Beyond." *Information Polity* 12 (1–2), 2007. (at 7)

Alongside this compromise there was some massaging of the language so that, in Kummer's terms,

> the wording of the final documents addresses the needs of both groups: it takes care of those governments trying to find their role in this new policy environment, and it respects the views of those who emphasize the importance of the private sector and civil society.[38]

Under these conditions, without reaching any kind of agreement about what Internet governance meant or who ought to be responsible for it, agreement on language on Internet governance for the Geneva Declaration of Principles and Plan of Action was reached late in the night of December 6, 2003, four days before the opening of the Geneva Summit and three days before a compromise would be reached on the creation of a funding program for the alleviation of the digital divide, the final unresolved Phase I issue.

Internet governance is discussed in the final Geneva Declaration of Principles in the following terms:

> 48. The Internet has evolved into a global facility available to the public and its governance should constitute a core issue of the Information Society agenda. The international management of the Internet should be multilateral, transparent and democratic, with the full involvement of governments, the private sector, civil society and international organizations. It should ensure an equitable distribution of resources, facilitate access for all and ensure a stable and secure functioning of the Internet, taking into account multilingualism.
>
> 49. The management of the Internet encompasses both technical and public policy issues and should involve all stakeholders and relevant intergovernmental and international organizations. In this respect it is recognized that:
>
> Policy authority for Internet-related public policy issues is the sovereign right of States. They have rights and responsibilities for international Internet-related public policy issues;
>
> The private sector has had and should continue to have an important role in the development of the Internet, both in the technical and economic fields;
>
> Civil society has also played an important role on Internet matters, especially at community level, and should continue to play such a role;
>
> Intergovernmental organizations have had and should continue to have a facilitating role in the coordination of Internet-related public policy issues;

---

38 Markus Kummer, "Agree to Disagree: The Birth of the Working Group on Internet Governance." in Daniel Stauffacher and Wolfgang Kleinwächter (eds.), *The World Summit on the Information Society: Moving from the Past into the Future*. New York: UN ICT Task Force, 2005 (at 246).

International organizations have also had and should continue to have an important role in the development of Internet-related technical standards and relevant policies.

50. International Internet governance issues should be addressed in a coordinated manner. We ask the Secretary-General of the United Nations to set up a working group on Internet governance, in an open and inclusive process that ensures a mechanism for the full and active participation of governments, the private sector and civil society from both developing and developed countries, involving relevant intergovernmental and international organizations and forums, to investigate and make proposals for action, as appropriate, on the governance of Internet by 2005.[39]

And, the Geneva Plan of Action adds (at 13b):

[...] The group should, inter alia:

iii) develop a working definition of Internet governance;

iv) identify the public policy issues that are relevant to Internet governance;

v) develop a common understanding of the respective roles and responsibilities of governments, existing intergovernmental and international organizations and other forums as well as the private sector and civil society from both developing and developed countries;

vi) prepare a report on the results of this activity to be presented for consideration and appropriate action for the second phase of WSIS in Tunis in 2005.[40]

The Working Group on Internet Governance (WGIG)

"Before it would have been possible to find a solution," Kummer realized, "there needed to be a common understanding that there was a problem that needed to be resolved."[41] It could be said that Internet governance as a new field of regulation was invented at this point of the WSIS.

The creation of the Working Group on Internet Governance (WGIG) was the response to the definitional problem. The WGIG secretariat was formed in July 2004 and held a series of open consultations in Geneva from 20-21 September

---

39 WSIS Executive Secretariat, *Geneva Declaration of Principles* (WSIS-03/GENEVA/DOC/4-E). (December 12, 2003). http://www.itu.int/wsis/documents/doc_multi.asp?lang=en&id =1161|1160

40 WSIS Executive Secretariat, *Geneva Plan of Action* (WSIS-03/GENEVA/DOC/5-E). (December 12, 2003). http://www.itu.int/wsis/docs/geneva/official/poa.html

41 Markus Kummer, "The Debate on Internet Governance: From Geneva to Tunis and Beyond." *Information Polity* 12 (1-2), 2007 (at 6).

2004. The composition of the Working Group was announced on November 11, 2004. There were 40 members drawn from civil society, governments, industry, the Internet technical community, NGOs and academia.[42] "All members of the group had," MacLean's "Brief History of WGIG" explains, "expertise in some aspect of Internet governance" and "many had also been involved in WSIS I and previous multi-stakeholder policy processes such as the G8 Digital Opportunities Task Force and the United Nations Information and Communication Technologies Task Force."[43] The usual UN overtures were made toward ensuring geographic and gender balance and toward promoting the representation of developing countries. Special advisor to the UN Secretary-General Nitin Desai was appointed chairman and Markus Kummer was named to direct the secretariat.

The process of nominating civil society representatives for the WGIG was frantic and controversial. Members of the WSIS-CS Internet Governance Caucus (IGC) took the initiative of positioning the IGC as the CS focal point for the WGIG nominations. The link between the IGC and the WGIG was a logical one and, with the CS Division closed and the Bureau and Plenary in the processes of reorganization (as discussed in Chapters 2 and 3), the move was endorsed as a pragmatic necessity by some of the more influential voices in broader WSIS CS. The ad-hoc process that was devised and refined on the fly as tight deadlines approached involved the following steps:

- all CS caucuses and working groups were invited to submitted a slate of up to 3 nominees for the WGIG to the IGC;
- the IGC assumed responsibility for refining the list down to 10–20 people and passing it along to the WGIG secretariat;
- in addition, the IGC would itself propose candidates for nomination;
- CS actors in all other caucuses were invited to join the IGC email list to participate in the process;
- any caucuses who did not wish to participate in the IGC process or who objected to its results were invited to submit their own nominations directly to the WGIG secretariat.[44]

---

42 For a list of WGIG membership see WGIG, *Report of the Working Group on Internet Governance*. (June 2005). http://www.wgig.org/docs/WGIGREPORT.doc

43 Don MacLean, "A Brief History of the WGIG." in William Drake (ed.), *Reforming Internet Governance: Perspectives from the Working Group on Internet Governance (WGIG)*. New York: UN ICT Task Force, 2005 (at 11).

44 Details on the processes distilled from CS email communication. See, in particular, Adam Peake, *Nominations: Working Group on Internet Governance (WGIG)*. [WSIS CS-Plenary]. (September 7, 2004).

The invitation to wider WSIS CS to participate in the IGC nominating process established the following criteria for CS WGIG candidates:

> We consider it critical that a balanced WGIG be drawn from a multi-dimensional consideration of diversity. That is, diversity in terms of sector, region, gender, and language background, among others, must be considered in assembling the WGIG. It is also important that there be a balance between members from developing and developed countries. It is also considered very important that candidates have a degree of knowledge of the issues, including policy, legal and technical, involved in the Internet governance debate. We also suggest candidates should have experience working in an international committee environment, be aware of ICT for development issues and human rights. No candidate is expected to have all these qualities, but we are suggesting they should be people with broad experience.[45]

The process through which the IGC would assess and make determinations about which names to include in the list forwarded to the WGIG secretariat was, as of the time the invitation was sent out, still being worked out. The invitation itself conceded that "There is not agreement for this course of action among members of the IG Caucus, but time is pressing and we feel WSIS Civil Society must have an opportunity to participate in this process."[46]

Some on the IGC listserv had voiced concerns that the invitation had been sent to wider CS prematurely, before all of the concerns expressed had been addressed and that consensus on this course of action had been demonstrated. Predictably, there were objections from civil society actors not involved in the IGC about the move. These tensions were only imported to and amplified by the debate that ensued on the IGC listserv about what sort of process should be established to evaluate the nominations. Opinions were divided, consensus seemed elusive and deadlines approached.

One of the IGC's coordinators took steps to organize a nominating committee of IGC members. This committee evaluated the 35 candidates that were proposed to the IGC from across civil society and forwarded 10 nominations to the WGIG secretariat. The Nominating Committee also backed a handful of civil society participants to serve as "connectors," members of various caucuses who would work to facilitate interaction with various thematic constituencies as the process expands to a broader range of issues.[47]

Certain WSIS-CS caucuses, for example the African Caucus, chose to nominate their own members directly to the WGIG secretariat and other organizations external to WSIS CS, such as the ICANN Non Commerical Users

---

45 Adam Peake, *Nominations: Working Group on Internet Governance (WGIG)*. [WSIS CS-Plenary]. (September 7, 2004).
46 Ibid.
47 Bertrand de La Chapelle, *Proposition de candidats pour le WGIG*. [governance]. (October 4, 2004).

Constituency (NCUC), also proposed candidates. The process through which these nominations were made and the role of the IGC would cause tension within CS. For example, according to IG Caucus member Milton Mueller:

> The IGC's method of selecting candidates was not a thing of beauty. Let's face the facts about that. It was rushed, disorganized, improvised, not transparent and in the end based far too much on personal connections and reflective of personal agendas.
>
> Given our institutional limitations it could not have been done much better. But let's not rationalize our failings, let's accept them as a basis for building better structures going forward. The NCUC [Non Commercial Users Constituency (of ICANN)] for example is a CS coalition with a charter and selected officers; its process for selecting nominees was far smoother, more transparent, and ultimately fairer, although it did not have to incorporate such a large and heterogeneous group into its deliberations.
>
> The caucus's process gave everyone involved some voice in the outcome. Let's hold our noses and accept the results and think more pro-actively about how to do it better next time. I would urge the critics of the results to also reflect on what is accomplished by picking at individual names at this juncture. Probably very little—the names have been transmitted and any attempt to change them raises more problems than it solves. We had our process, now let's live with it.[48]

Other civil society members were more pointed in their criticism of the selection process:

> One example of how the structure and the organization of CS has had a major influence on the content and political actions that have been produced is the selection of CS delegates who were to form the Working Group on Internet Governance (WGIG). Out of ten CS members, at least three represented the same organization, while several delegates were admitted even though they normally should have entered as part of the private sector. The selection process of the WGIG is without any doubt one of the clearest examples of the hidden pitfalls of the MSP [multi-stakeholder partnership] model.[49]

Regardless, in the end, virtually all of the CS nominees were included in the WGIG.

The first formal meeting of the WGIG was convened on November 23, 2004. The WGIG held four meetings: November 23-25, 2004; February 14-18, 2005; April 18-0, 2005; and June 14-17, 2005. All of these meetings took place at UN offices in Geneva although, during the final meeting, the group decamped to the Chateau de Bossey located in the countryside environs of Geneva to facilitate the report drafting process. In addition to the open consultations, a variety of WSIS

---

48   Milton Mueller, *Recommendations for WGIG*. [WSIS CS-Plenary]. (October 7, 2010).
49   Beatriz Busaniche, "Civil Society in the Carousel: Who Wins, Who Loses and Who is Forgotten by the Multi-stakeholder Approach?" In Olga Drossou and Heike Jensen (eds.), *Vision In Process*. (2005). http://www.worldSummit2005.de/download_en/Visions-in-ProcessII (1).pdf

regional and sub-regional meetings provided input to the WGIG's work. ICANN, the ISOC and a variety of Internet technical organizations and academic institutions also held what were described by WGIG member Bill Drake as "various contributory sessions to the ongoing debate."[50]

The WGIG submitted a preliminary report to WSIS Phase II, PrepCom II which was discussed in a plenary session on February 24, 2005, and released its final report July 14, 2005. The methodology of the WGIG was to start by identifying all of the public policy issues relevant to Internet governance and then to progressively build bottom-up toward a working definition of Internet governance that would, according to WGIG member Don MacLean, "capture the essential elements that were common to all of these issues."[51] This process changed the common understanding of Internet governance on the international scene.

From ICANN to Internet Governance

The WGIG report provides an accounting of what were determined to be the 13 highest priority "public policy issues that are potentially relevant to Internet governance,"[52] which were organized into four "key public policy areas:"

- issues relating to infrastructure and the management of crucial Internet resources;
- issues relating to the use of the Internet;
- issues relevant to the Internet but, like intellectual property rights or international trade, having "an impact much wider than the Internet and for which existing organizations are responsible";
- issues related to development and capacity building in developing countries.[53]

---

50 William Drake (ed.), *Reforming Internet Governance: Perspectives from the Working Group on Internet Governance (WGIG)*. New York: UN ICT Task Force, 2005 (at 4).
51 Don MacLean, "A Brief History of the WGIG." In William Drake (ed.), *Reforming Internet Governance: Perspectives from the Working Group on Internet Governance (WGIG)*. New York: UN ICT Task Force, 2005 (at 11).
52 Namely: Administration of the root zone files and system; interconnection costs; Internet stability, security and cybercrime; spam; meaningful participation in global policy development; capacity-building; allocation of domain names; IP addressing; intellectual property rights (IPR); freedom of expression; data protection and privacy rights; consumer rights; multilingualism. See WGIG, *Report of the Working Group on Internet Governance*. (June 2005) (at 4–6) http://www.wgig.org/WGIG-Report.html
53 WGIG, *Report of the Working Group on Internet Governance*. (June 2005). http://www.wgig.org/docs/WGIGREPORT.doc

The WGIG accepted that if a policy field of Internet governance were going to be defined, then issues outside the mandate of ICANN had to be a part of it. Furthermore, the WGIG was clear that, if issues that were external to ICANN could be considered to be Internet governance issues, then Internet governance had to be defined as something broader than the management of the DNS system. It is significant in this respect, that within the WGIG's efforts to map the policy field of Internet governance, "administration of the root zone files and system" was but one of the thirteen policy issues. It was also only one of a series of issues mentioned under the "Issues relating to infrastructure and the management of crucial Internet resources" policy area.[54]

As promised, the WGIG report presented a working definition of Internet governance:

> Internet governance is the development and application by Governments, the private sector and civil society, in their respective roles, of shared principles, norms, rules, decision-making procedures, and programmes that shape the evolution and use of the Internet.[55]

The WGIG report provided an accounting of the roles and responsibilities of governments, civil society and the private sector. It concluded that "some adjustments needed to be made to bring" the existing Internet governance arrangements "more in line with the WSIS criteria of transparency, accountability, multilateralism and the need to address all public policy issues related to Internet governance in a coordinated manner." In response it presented a proposal for creation of a forum that would function as a "new space for dialogue for all stakeholders on an equal footing on all Internet governance related issues."[56] The WGIG report also proposed four different institutional models that could serve as the basis of a reformed system of "global public policy and oversight." These are outlined in Table 3.[57]

---

54 Ibid.
55 Ibid. (at 3).
56 Ibid. (at 9).
57 *Italics*: Proposed new organization/institution; Underscored: Existing organizations/institutions subject to reform (reform details in brackets).

	Oversight	Government Input	Policy making	Management of Root	Buoyed	Sidelined
Model 1	Global Internet Council (GIC)	GIC	GIC	ICANN, (formally accountable to GIC)	-UN -govs.	-ICANN GAC -US DOC -Civil Society and Private Sector (advisory roles)
Model 2	None	ICANN GAC (enhanced)	IGF* [coordination / analysis function]	ICANN	-govs. (only slightly)	-US DOC -UN
Model 3	International Internet Council (IIC)	IIC	IIC	ICANN, (accountable to IIC, with a host country agreement)	-govs.	-ICANN GAC -US DOC -UN -Civil Society and Private Sector (advisory roles in IIC)
Model 4	Oversight Committee (Appointed by the Global Internet Policy Council [GIPC])	-GIPC -Oversight Committee -Advisory function in WICANN (rebaptised ICANN)	GIPC GIGF (Global Internet governance forum)	WICANN ("World ICANN": linked to UN with a host country agreement)	-UN -govs.	-US DOC -Civil Society (observer role in GIPC and WICANN in observer capacity) -Private Sector (lead operational role in WICANN but observer status in GIPC and WICANN policy making)

Table 3: Summary—The WGIG Models for Institutional Reform of Global Internet Governance

When the WSIS discussion of Internet governance resumed at Phase II, PrepCom III, there was palpable appreciation evident for the work of the WGIG direction and membership. For staunch supporters of multi-stakeholder global governance, the WGIG is often pointed to as an ideal; a common refrain amongst WSIS civil society participants who participate in the IGF is that "The IGF is no WGIG." Former WGIG member Bill Drake makes the case that the WGIG facilitated the WSIS negotiations on Internet governance by:

- providing a common vocabulary to the terms and issues being debated

that reduced the frequency of, and frustration over, the sort of misunderstandings that had plagued the Internet governance debates during WSIS Phase I;
- allowing for civil society, Internet technical and industry experts to be in the room while government representatives addressed Internet governance issues, thereby facilitating a process of institutional learning that organically cleared up and filled in many of the misconceptions and knowledge gaps that had been evident during WSIS Phase I (and thus also demonstrating the benefits of multi-stakeholder collaboration);
- creating a non-binding process wherein members could discuss openly and thus clarify not only the issues themselves but where various other delegations stood on them;
- incubating and developing the concept of an Internet governance forum;
- deflecting the calls for reform away from a focus on the ITU's possible role in global Internet governance;
- creating a year and a half long period of détente while the WSIS process largely put aside the issue of Internet governance waiting for the conclusion of the WGIG, thus allowing for the temperature to be reduced a bit and for delegations to better coordinate and work through their positions;
- working through concepts and definitions.[58]

As we have seen, the WSIS did not step into a vacuum on Internet governance. The context in which the issue was raised at the WSIS was set by a long history of controversy over Internet domain names and numbering[59] as well as the conventional view that pretty much everything else related to the Internet was some combination of inherently democratic and immune to control.[60] The result, according to Drake, was that when the term first emerged during the WSIS

> the nearly standard practice [had] been to equate the term 'Internet governance' with the social organization of Internet identifiers and the root server system and, by extension, the functions performed by the ICANN.

This "narrow definition," Drake continues,

---

58 See William Drake (ed.), *Reforming Internet Governance: Perspectives from the Working Group on Internet Governance* (WGIG). New York: UN ICT Task Force, 2005.
59 See Milton Mueller, *Ruling the Root: Internet Governance and the Taming of Cyberspace*. Cambridge, Mass: MIT Press, 2002.
60 See, for example, Jack L. Goldsmith and Tim Wu, *Who Controls the Internet? Illusions of a Borderless World*. New York: Oxford University Press, 2006; and Darin Barney, *Prometheus Wired: The Hope for Democracy in the Age of Network Technology*. Vancouver: UBC Press, 2000.

was inconsistent with the empirical reality that there are a variety of collectively applicable, private and public sector rules, procedures and programs that shape both the Internet's infrastructure (physical and logical) and the transactions and content conveyed thereby.

What emerged in its place over the course of the first phase of the WSIS was what Drake calls "a broader and more holistic conception that could encompass the full range of Internet governance mechanisms and facilitate their systematic evaluation and coordinated improvement."[61]

Explicitly disaggregated from the question of ICANN oversight, certain public policy issues that had been discussed over the first phase of the WSIS lent themselves to relatively uncontroversial resolution in Phase II. This was the case, for example, with international interconnection costs, the questions related to the regional Internet registrars and the linkages between government sovereignty and CCTLDs. Other issues that had been prominent topics of discussion during the first phase of the WSIS and that had been identified by the WGIG as legitimate concerns within the field of Internet governance, such as free software and freedom of expression/communication rights, virtually dropped off the WSIS agenda altogether.

But the question of ICANN oversight, despite the work that the WGIG did to strip it of the mostly external public issues that had been grafted onto it over the course of Phase I, remained a contested and controversial focus throughout.

The US government, for instance, released a statement of four Internet principles in June of 2005 and conceded that

> Governments have legitimate interest in the management of their country code top level domains (CCTLD). The United States recognizes that governments have legitimate public policy and sovereignty concerns with respect to the management of their CCTLD. As such, the United States is committed to working with the international community to address these concerns, bearing in mind the fundamental need to ensure stability and security of the Internet's DNS [...] we encourage an ongoing dialogue with all stakeholders around the world in the various fora as a way to facilitate discussion and to advance our shared interest in the ongoing robustness and dynamism of the Internet.

However, the US also insisted that

> ICANN is the appropriate technical manager of the Internet DNS [...] the United States is committed to taking no action that would have the potential to adversely im-

---

61 William Drake, "Reforming Internet Governance: Fifteen Baseline Propositions." in Don MacLean (ed.), *Internet Governance: A Grand Collaboration*. New York: UN ICT Task Force, 2004 (at 144).

pact the effective and efficient operation of the DNS and will therefore maintain its historic role in authorizing changes or modifications to the authoritative root zone file.[62]

But, despite making progress in achieving broad-based recognition—even from the US government—of the need to establish public policies related to certain Internet governance issues external to the question of ICANN's oversight, many governments still wanted control of the DNS switch.

Internet Governance at WSIS Round III: The Tunis Compromise

Responses to the WGIG report were collected over the course of the summer of 2005[63] and the effort to negotiate an agreement on global Internet governance resumed at the WSIS with the opening of the third and final planned PrepCom of the second phase on September 19, 2005. As during the run up to the first phase, it was clear that it was going to be difficult—if not impossible—to reach an agreement before the summit. Whereas Internet governance was just one of a handful of issues that were not close to resolution by the final PrepCom of Phase I, PrepCom III of the second phase consisted of two subcommittees: Subcommittee A devoted to negotiation of Internet governance and Subcommittee B that was devoted to negotiation of everything else to do with the second phase of the summit. The debate over Internet governance had grown in significance at the WSIS, but had it evolved?

The work of the WGIG in itemizing and explaining Internet governance issues undoubtedly proved instructive to many delegations and provided a lexicon for framing perspectives going forward. While such efforts may have cleared up previous misunderstandings and prevented new ones from occurring, the WGIG's terms of reference provided a common language in which stakeholders could express not only their agreements, but their differences as well.

The responses to the WGIG report revealed a general consensus that the WGIG definition of Internet governance was at least workable and that there was little objection to using the WSIS to express aspirations that a handful of

---

62   US Department of Commerce, National Telecommunications and Information Administration (NTIA), *Domain Names: U.S. Principles on the Internet's Domain Name and Addressing System*. http://www.ntia.doc.gov/ntiahome/domainname/usdnsprinciples_06302005.htm

63   WSIS Executive Secretariat, *Compilation of Comments Received on the Report of the WGIG*. (WSIS-II/PC-3/DT/7(Rev. 2) E). (September 23, 2005). http://www.itu.int/wsis/docs2/pc3/working/dt7rev2.doc

the public policy goals outlined in the WGIG report—combating spam, increasing capacity of developing countries to participate in Internet governance, reinforcing the principle that governments have sovereignty over their CCTLDs, underlining the need for multilingualism, etc.—might eventually be realized.[64] In regard to institutional and operational questions—about the roles and responsibilities of different stakeholders, about the creation of an Internet governance forum and crucially about the question of governmental oversight—however, the comments on the WGIG report revealed stark differences of opinion.

These differences were emphatically underlined only minutes into the first session of Subcommittee A of PrepCom III on September 20, 2005, by a Brazilian intervention that described Internet governance in "three words: lack of legitimacy." From there, Brazil moved on to argue that the adage "if it ain't broke, don't fix it" was a non-sensical and Orwellian construct, and to suggest instead that delegates consider "Stein's law" that says that "things that can't go on forever don't."[65] Brazil, in other words, got straight to the point in summarizing the veracity of the calls for reform of global Internet governance and in making an effort to disavow WSIS participants of any notion that the WGIG had somehow eliminated the fundamental differences that existed between the status quo and reform perspectives during the first phase.

What was clear from the comments that were received on the WGIG report and from the early discussions at PrepCom III was that the difference in opinion, though it had not gone away, had evolved. There was little direct discussion of the ITU in the PrepCom III round of the debate. Instead of ICANN vs. ITU, the discussion was largely over "the current system" vs. "a different system with a larger role for governments." A knock-on effect of this discursive shift was the space that it created between the two poles for middle ground positions. Whereas many delegations had previously been unsure whether they preferred the ICANN or the ITU, or had joined one or the other camp as a lesser of two evils choice, over the course of PrepCom III the notion that change to the existing system of global Internet governance could

---

64 Ibid.
65 The "if it ain't broke, don't fix it" line of defense of the status quo was articulated in various formations on numerous occasions. For example, it is spelled out in exactly those terms in then ICANN executive and 'Internet founding father' Vinton Cerf's contribution to a UN-ICT Task Force published volume of articles on Internet governance circulated to WSIS participants, Daniel Stauffacher & Wolfgang Kleinwächter (eds.), *The World Summit on the Information Society: Moving from the Past Into the Future*. New York: United Nations Information and Communication Technology Task Force, 2005

occur in degrees and was not an all-or-nothing proposition emerged as a viable negotiating position.

For their part, certain civil society actors made efforts to help WSIS participants understand these tensions and the compromise solutions that were on the table. On separate days over the course of Phase II, PrepCom III, the IGC organized information sessions on the Internet Governance Forum that had been proposed by the WGIG report and on the notion of Internet governance oversight. These events were structured around presentations from academic experts and experienced practitioners from civil society, many of whom had been members of the WGIG, followed by questions from audience members and further discussion. The primary objective of these sessions was facilitative: they were intended as informational resources in the hope that more detailed explanations of some of the concepts being debated at WSIS might help clear up any misconceptions hindering negotiation progress. There was an undeniable element of proselytizing to them as well, in particular where they concerned enthusiasm for the Internet Governance Forum as a compromise to the negotiation deadlock. The idea of the forum was largely claimed (by CS delegates at least) as the brainchild of civil society members of the WGIG. Its approval by the WSIS would thus reflect positively on both individual civil society members closely associated to it and validate the role of civil society within global policy development institutions such as the WGIG. The establishment of the Internet governance forum would also carve out a formally recognized role for civil society in the post-Tunis institutional architecture of Internet governance.

By the end of a first week in which the meetings of Subcommittee A were largely devoted to general discussion of the issues and repeated debate of procedural concerns (including the status of various working documents, the participation of non-governmental stakeholders and, generally, the working methods of the subcommittee), a "Chair's Paper" draft of the text for the Tunis final documents on Internet governance was prepared.[66] The level of detail in most sections reflected a growing sense of confidence on the part of the chair that consensus was emerging on certain issues. But the part labelled "Follow-up and Possible Future Arrangements" contained only a point-form laundry list of the really contentious issues: oversight, institutions and the creation of an Internet governance forum. The comments that were received on this part of the first draft made it clear that the reform-minded governments and the defenders of the status quo were still very far apart.

---

66 Chair of the Sub-Committee A (Internet Governance), *Chapter Three: Internet Governance Chair's paper* (WSIS-II/PC-3/DT/10-E). (September 23, 2005). http://www.itu.int/wsis/docs2/pc3/working/dt10.doc

The EU and a "New Cooperation Model"

A significant change in the dynamics of the WSIS IG negotiations occurred on Wednesday September 28, 2005.

After arranging a meeting of senior officials—many of whom flew to Geneva from capitals all around Europe just for the occasion—the European Union introduced its own proposal for follow-up and future arrangements. The proposal was introduced by the head of the British delegation (acting in his capacity as head of the EU delegation) as "something that we hope the people from the extreme positions of the discussion could come to agree on." "We hope," he continued—obviously anticipating how it might actually be received—"that they will take it away and react tonight or react tomorrow rather than reacting as they hear it or read it" and, "with that explanation...," he read aloud a proposal from the EU for "a new cooperation model" for global Internet governance that involved:

- international government involvement at the level of principles over various naming, numbering and addressing related matters including: allocation of IP blocks; procedures for changing the root zone file (particularly for new top level domain name creation and changes to CCTLD managers); DNS system rules; contingency planning for ensuring the continuity of the DNS functions; establishment of arbitration and dispute resolution mechanisms linked to international law;
- creation of an Internet governance forum; and, in parallel, a separate process to transition to the new model of international cooperation.

The "new cooperation model" was to be based on the principles of not replacing existing mechanisms or institutions, maintaining a multi-stakeholder public private partnership, and reinforcing the involvement of government in the "principal issues of public policy." The latter objective was to be accomplished, the proposal suggested, without granting governments any "involvement in the day-to-day operation" or threatening the existing "architectural principles of the Internet, including the interoperability, openness and the end-to-end principle." [67]

Despite being introduced with a disclaimer that called for other delegations to sleep on the proposal before reacting, the response of the Americans suggested that they immediately interpreted the proposal as a challenge to

---

67 European Union, *Proposal for Addition to Chair's Paper Sub-Com A Internet Governance on Paragraph 5 "Follow-up and Possible Arrangements.* (September 28, 2005). http://www.itu.int/wsis/docs2/pc3/contributions/sca/EU-28.doc

their authority. At the conclusion of the evening sitting of Subcommittee A on September 28, 2005, the US delegation asked for and was granted the floor. The interventions of US spokesperson Dick Beaird had, over the course of PrepCom III, offered a master-class in the rhetoric of diplomacy. Disarmingly civil, Beaird's interventions generally managed to be non-confrontational and support points made by other delegations while deflecting focus and discussion away from key issues and the specifics of US positions on them. In response to the EU proposal, rather than deflecting attention away from the issue, the US delegation chose to be direct and to the point in emphatically restating its position and what it was not willing to accept. "We want to make perfectly clear once again," Beaird began

> [the distinction] between public policy and the day-to-day operations of the Internet. The day-to-day operations of the Internet, of which any changes or modifications to the authoritative root zone file is a part, is essential to the trust and confidence that the world may have and should have in the Internet. It is a responsibility that the US takes with great seriousness and we will not do anything to adversely impact that responsibility. On the other hand, there are many issues that we would say fall in the domain of the public policy realm. That includes: spam, viruses, cybersecutiry, cybercrime, all of the issues that we are very much concerned with and that we wish to engage in actively on a dialogue that will lead to the resolution of those issues.

Concluding with the salutation, "Mr. Chairman, these are issues that this delegation takes as fundamental,"[68] the Americans emphatically reinforced the point that this was—in no uncertain terms—their red line. The WSIS would either reach an agreement on Internet governance that did not challenge it, or would reach no agreement that the Americans would accept.

The US was not alone in arriving at this interpretation of the EU proposal. Over the course of the first session of Subcommittee A on Thursday September 29, 2005, a series of delegations took the floor to express interest in and support for the EU proposal. Making matters worse for EU/American relations, these new-found friends included the governments who had been most vocal in their calls for reform to the global Internet governance system over the course of the WSIS, such as Brazil and China; governments such as Saudi Arabia whose interest in communication regulation has historically been very different from that of most European countries; and countries such as Iran, Venezuela and Cuba with whom the United States was actively engaged in diplomatic hostilities of varying degrees at the time.

---

68   ITU, *Broadcasting Services for the Third Meeting of the Preparatory Committee for WSIS (PrepCom-3)*. http://www.itu.int/ibs/WSIS/p2/pc3/index.phtml

Delegations supporting proposals from Iran, Brazil and Argentina were encouraged by the chair to discuss their positions with the EU with an eye to condensing the series of proposals into one. The reports returned the next day—what was supposed to be the final day of PrepCom III—were clear that common ground had not been found between the EU and the other delegations. Argentina, Iran and Russia were among the delegations that, after discussing options with the EU, decided that their positions were distinct enough from those of the EU to warrant separate proposals. Brazil went as far as submitting a paper outlining the modifications they would have required in order for the EU proposal to be supported by their delegations. To which the EU responded, "we do not think that we could sign-up to the proposals brought back."

The consideration and eventual rejection of the original EU proposal by the more reform-minded governments bought the EU more time to think through some of the vaguely worded elements of the draft. There had been a sudden spike in media attention on the WSIS that included a series of "EU and US clash over control of net" headlined stories in papers such as the *New York Times* and the *International Herald Tribune* on September 30, 2005,[69] implying that the EU might be capitalizing on anti-Iraq War backlash to send a message about US unilateralism. The push from Iran, Brazil and others to get the EU to say even more about oversight of ICANN also presented the EU with the opportunity to distance itself from the politically undesirable company it found itself in and spin the proposal as less of a departure from the American position as the negotiations moved forward.

With that, Subcommittee A of PrepCom III ended the way it had started: with a debate over the status of a document (the WGIG report in the beginning, a proposed Chair's paper in the end). A draft declaration that was coming together on many issues, but sparse on the crucial questions of institutional reform, as well as a series of proposals on oversight and the Internet governance forum, were forwarded for further negotiations planned in a PrepCom III resumed session scheduled for the days preceding the Tunis phase of the summit.

Of the proposals on the table, those of Saudi Arabia (on behalf of the Arab Group) and Iran centered on explicit creation of a new intergovernmental institution for oversight of the Internet. Proposals from the EU, Ghana (on behalf of the African Group), Argentina and Russia each implied some process of gradual internationalization of Internet governance. This was typically framed as an increase in the role of governments in Internet public policy that

---

69 Tom Wright, "EU and US Clash Over Control of Net." *The New York Times*. (September 30, 2005). http://www.nytimes.com/2005/09/29/business/worldbusiness/29iht-net.html

remained vague about where Internet public policy stops and oversight begins. The proposals from Brazil, Canada and Japan focused only on the creation of the forum, though Ghana, Argentina, the EU and Saudi Arabia also advocated creation of a forum. The Brazilian delegation framed its proposal for the forum as separate from its view on an intergovernmental oversight mechanism, while Canada did not support new oversight at all, and the Japanese proposal was clear that the discussion of new models should continue in the forum.[70]

In other words, the first sitting of PrepCom III concluded with calls for radical overhaul of the system that were entirely at odds with the line in the sand that had been drawn by supporters of the status quo. In the middle had emerged a perspective centered around progressive or evolutionary change and the idea that a forum could be created as a new institution whose non-binding mandate would not substantively impact the status quo. Faced with a similar conundrum as the first phase was winding down, the WGIG had been proposed as a mechanism for continuing the debate and providing a way out of an intractable difference of opinion. At the second phase, it was clear that the IGF could act as a similar type of way out.

Between the conclusion of the September sitting of PrepCom III and the resumed session in November 2005, work and politics continued behind the scenes. In the lead-up to the resumed PrepCom III session, the US is said to have exercised pressure at the highest diplomatic levels of its special friendship with the UK in order to convey the gravity with which it viewed the EU proposal.[71]

---

70 All of these proposals are available on the WSIS official website. http://www.itu.int/wsis/documents/listing-all.asp?lang=en&c_event=pc2|3&c_type=all|

71 In what has become the stuff of WSIS lore, Condoleezza Rice is rumored to have personally sent her British counterpart Jack Straw a diplomatic letter expressing American concern for the EU position on Internet governance (the British were, at the time, head of the EU delegation in their capacity as president of the EU). What is claimed to be a text of the letter found its way on to the Internet. See Kieren McCarthy, "Read the letter that won the internet governance battle: Condoleezza Rice's missive to the EU," *The Register*. (December 2, 2005). http://www.theregister.co.uk/2005/12/02/rice_eu_letter/. Published excerpts include:

"The success of the Internet lies in its inherently decentralized nature, with the most significant growth taking place at the outer edges of the network through innovative new applications and services. Burdensome, bureaucratic oversight is out of place in an Internet structure that has worked so well for many around the globe. We regret the recent positions on Internet governance (i.e., the "new cooperation model") offered by the European Union, the Presidency of which is currently held by the United Kingdom, seems to propose just that—a new structure of intergovernmental control over the Internet [...] we ask

The Tunis Compromise

Subcommittee A of PrepCom III resumed in Tunis on November 14, 2005, with a reminder from the chair that "we have a responsibility to citizens and constituencies around the world to come up with a result." In plenary sessions and break out drafting groups, debate continued, often picking up where it had left off in September.

On the forum function, the US was willing to admit that "the United States always believes in dialogue" but Australia was insisting that the forum should be "pro-market" and that it was "not the time to talk about governments." When Saudi Arabia requested insertion of "a new cooperation model" into the text, Australia was quick to point out that there was not agreement that WSIS should advocate new models and the US insisted that, rather than a middle ground position, "a new cooperation model has become indistinguishable with a new intergovernmental model."

From there, discussion shifted to the question of whether the ICANN Governmental Advisory Committee (GAC) could be reformed to improve the participation of governments or whether a new body needed to be formed instead. Arguing that "we will never find a compromise on the word oversight," the EU responded that the goal of the WSIS should come down to "creat[ing] a legal ground for the improvement of the GAC." At that point, the chair introduced a letter from then-chairman of the ICANN board Vinton Cerf to GAC chair Mohamed Sharil Tarmizi discussing the need for reform. Acknowledging that, through the WSIS, "a great deal of attention has been devoted to the role of governments in the process of 'Internet Governance,'" Cerf suggests scheduling a meeting to "discuss how to best address these concerns, and what measures need to be taken to make our cooperation more effective, including ensuring the participation of developing countries."[72] At a crucial moment, ICANN was making very public overtures to its critics.

At the start of the evening session on November 14, in response to a new chair's paper that included creation of a forum, discussion focused on the institutional teeth that the forum would be given. Australia argued against use of the word "governance" in the name of the forum, suggesting a change from "Internet Governance Forum" to "Internet Dialogue Forum." The same intervention also advocated the ISOC as the host organization of the forum, a call which was echoed by the US in an intervention that criticized the UN and the

---

the European Union to reconsider its new position on Internet governance and work together with us to bring the benefits of the Information Society to all."

72 ICANN, *Letter from ICANN Chairman Vint Cerf to GAC Chairman.* (November 9, 2005). http://www.circleid.com/posts/letter_from_icann_chairman_vint_cerf_to_gac_chairman/

ITU as unsuitable potential host organizations. Imagining newspaper headlines along the lines of "Internet Kindergarten Forum," Brazil, for one, voiced concern that a non-binding forum hosted outside of the UN system would not be taken seriously as en effort to internationalize Internet governance.

The final planned day of negotiations began with the EU's self-proclaimed effort to "fit between two opposing sides." They put forth a proposal that the forum be accompanied by a parallel process of "enhanced cooperation" which would "enable governments, on equal footing to carry out their roles and responsibilities in international public policy issues pertaining to the Internet not in the day to day or technical operation or arrangements."

The US was satisfied that, within the proposed compromise, "the UN does not have a regulatory or oversight function." Thus, it proved willing to accept a proposal to ask the Secretary-General of the UN to convene the IGF. To the US, Australia and others, the Internet Society (ISOC) would have been a preferable institutional home for the IGF. But the UNSG was a good deal less problematic as convener of the forum for these delegations than the ITU would have been. The vague parallel process of enhanced cooperation was accepted in principle. After resolving a debate over the relative merits of the words "framework" vs. "mechanisms" and of "structures" vs. "systems," the massaging of the final language concluded late on November 15, 2005. Saudi Arabia (as well as Iran and South Africa) threatened to reopen previously agreed upon paragraphs if the US did not concede to removing the clause "if justified" from the section on creating suitable multilateral mechanisms. The US promised to respond in kind by revisiting the concessions it had made elsewhere. The EU proposed reformulating "if justified" to "when justified," which was rejected by the Saudis, Iranians and South Africans who then accepted a follow-up EU proposal for "where justified." With that the WSIS negotiations on Internet governance concluded and the square brackets were removed around the Tunis texts on Internet governance.[73]

---

73 The Tunis *Agenda for the Information Society* and its companion *Tunis Commitment* are available online on the WSIS official website. http://www.itu.int/wsis/documents/doc_multi.asp?lang=en&id=2266|2267

• CHAPTER SIX •

Implementation and Follow-up

The WSIS as a Test for Implementation and Follow-up?

The World Summit on the Information Society was not only an experiment in multi-stakeholder participation; it was also a dry run for a new approach to coordinating the implementation and follow-up process of a UN Summit.

In parallel to the first phase of the WSIS, the UN formed an Ad Hoc Working Group of the General Assembly to examine the "integrated and coordinated implementation of and follow-up to the outcomes of the major United Nations conferences and summits in the economic and social fields." The report of this Working Group determined that "progress in implementation has been insufficient and therefore the time has come to vigorously pursue effective implementation." In response, UN General Assembly resolution 57/270 B was passed in July of 2003, emphasizing that

> the United Nations system has an important responsibility to assist Governments to stay fully engaged in the follow-up to and implementation of agreements and commitments reached at the major United Nations conferences and Summits, and invites its intergovernmental bodies to further promote the implementation of the outcomes of the major United Nations conferences and Summits (57/270 B at para 6).[1]

The mid-WSIS adoption of resolution 57/270 B in 2003 by the General Assembly of the United Nations marked a change in the UN system's approach to implementation, follow-up, and monitoring of the decisions taken at larger conferences and international summits. Going forward, the United Nations sought to ensure that resources and energy invested in such events would materialize into concrete initiatives. The adoption of demonstrable and systematic follow-up and implementation plans would, it was felt, reinforce the credibility of large, expensive and time-consuming international events. The

---

1 United Nations General Assembly, *Resolution A/RES/57/270 B: Integrated and Coordinated Implementation of and Follow-up to the Outcomes of the Major United Nations Conferences and Summits in the Economic and Social Fields.* (July 3, 2003). http:// www.unctad.org/en/docs/ ares57270b_en.pdf

WSIS was the first summit to take place after the adoption of resolution 57/270 B.[2]

With the first phase of the WSIS entering its final stages as this resolution was being passed, the two-phased structure of the WSIS presented itself as an opportunity to put principle into practice. Building plans for its own implementation and follow-up into the outcomes of the summit itself emerged as a politically shrewd not to mention logical and important agenda item for Phase II.

The first phase of the summit was conventional with respect to implementation and follow-up issues. The Declaration of Principles that presented the political and ethical foundations of the summit as well as the joint vision for an international social and political project was adopted in 2003. The Geneva Plan of Action was adopted hastily, following an arduous negotiation process. The Plan of Action officially translated the vision described in the Declaration of Principles into strategies for addressing the problems raised at the summit by proposing more or less clearly defined initiatives that were either completely or partially non-binding. No meaningful mechanisms were set up during Phase I to ensure that the decisions adopted and the goals set in Geneva would be effectively implemented. Furthermore, no strategy for following up on the initiatives that resulted from the summit was formalized either. At best, the Geneva Plan of Action proposed a series of actions to be undertaken by the summit's many participants and encouraged the creation of indicators to measure progress on the issues that had been raised. Phase II of the WSIS was expected to fill in some details around these vague ambitions.

Seeking to validate the Tunis phase as more than simply a rehashing and continuation of previously held (and largely unproductive) debates over Internet governance, digital divide financing and other issues left unresolved by Phase I, summit organizers presented the Tunis Summit as a "summit of solutions," insisting that the second phase of the WSIS had its own role to play in the sense that it would go beyond the traditional framework of UN Summits by agreeing to and articulating concrete follow-up and implementation mechanisms.

---

2   Resolution A/RES/57/270 B notably mentions: the implementation of decisions adopted during large UN Summits to be unsatisfactory, the role of the United Nations in assisting governments in their implementation actions, and the need to set up concrete measurements for evaluating the implementation and follow-up of the decisions adopted at large international events that take place under the UN system. See United Nations General Assembly, *Resolution A/RES/57/270 B: Integrated and Coordinated Implementation of and Follow-up to the Outcomes of the Major United Nations Conferences and Summits in the Economic and Social Fields*. (July 3, 2003). http://www.unctad.org/en/docs/ares57270b_en.pdf

The implementation and follow-up discussion was effectively launched in August 2004 with the convening of a stocktaking process. This meeting followed on a decision taken during the first meeting of the Phase II preparatory committee, agreeing to the following objective for the Tunis phase:

> Follow-up and implementation of the Geneva Declaration of Principles and Plan of Action by stakeholders at national, regional and international levels, with particular attention to the challenges facing the Least Developed Countries.[3]

A group of summit actors from different constituent groups gathered at this meeting, which essentially began the discussions and dialogue on the issue of the implementation and follow-up of Geneva decisions. The meeting also emphasized the need to establish measurable indicators and evaluate the contribution of the different parties working to reduce the problems identified at the WSIS.

> The stocktaking for WSIS implementation presents particular challenges because of the multi-stakeholder nature of the process, and because the information society overlaps the mandates of several different UN organisations. Nevertheless, there is the potential for the stocktaking to make a major contribution to coordination and harmonisation of efforts in this area, and to reduce the potential duplication of work. WSIS should set the lead within the UN system in using ICT tools effectively in communicating its message.[4]

The desire to assign tasks and responsibilities to the different parties taking part in the summit resonated with the Secretary-General of the ITU, who sent a letter to the governments, international organizations, members of the private sector, and civil society actors participating in the WSIS on October 4, 2004. The letter stated that the Executive Secretariat of the WSIS was working on compiling a list of the activities undertaken by different stakeholders in order to implement the Geneva Plan of Action, and requested every participant to fill out a questionnaire that would identify and outline their fields of activity as well as the actions they had taken thus far. The questionnaire was based on the 11 different themes defined in the Geneva Plan of Action.[5]

---

3   WSIS Executive Secretariat, *Final Report of the Preparatory Meeting* (PrepCom-1 of the Tunis phase) (WSIS-II/PC-1/DOC/06). (June 26, 2004). http://www.itu.int/wsis/docs2/pc1/doc6.doc
4   WSIS Executive Secretariat, *WSIS Stocktaking: Proposed Format and Approach Discussion Paper.* (August 2004). http://www.worldSummit2005.de/download_en/Proposed-approach-for-stocktaking-August-2004.pdf
5   WSIS Executive Secretariat, *Geneva Plan of Action* (WSIS-03/GENEVA/DOC/5-E). (December 12, 2003). http://www.itu.int/wsis/docs/geneva/official/poa.html

The goal of this initiative was to create a database that would synthesize the different WSIS-relevant activities being undertaken, thereby identifying the fields that required more work, as well as to contribute to gauging any progress that may have occurred since the end of the first phase in 2003. The results obtained would feed into the negotiation process at PrepCom II of Phase II.

Civil Society's Responsibilities and Challenges

Implementation and follow-up issues were generally twofold challenges for civil society at the WSIS. First, civil society sought to ensure that the WSIS adopted implementation and follow-up mechanisms that would transform the decisions adopted in Geneva and Tunis into initiatives that would be appropriate to the task of effectively addressing identified problems and concretizing the vision of the summit. This involved four parts:

1. specific mechanisms for implementation and follow-up measures had to be determined;
2. tasks and responsibilities had to be determined and divided among the different actors;
3. the level that would be focused on (international, regional, or national) and the specific measures that would be applied to each level had to be determined. The responsibilities behind each measure had to be divided amongst the different stakeholders;
4. the international bodies that would head the implementation and follow-up mechanisms, coordinate activities, and ensure that the implementation and follow-up process ran smoothly had to be determined.

The task was complex in and of itself and continued to be debated until very late into the preparatory process. A consensus on these issues was not reached until the third PrepCom of the Tunis phase.

Second, civil society organizations refused to be excluded from the implementation and follow-up mechanisms themselves given that such arrangements represented the logical venue for building on the resources, time, and energy that had been invested in the WSIS over four years. Since the official role played by civil society in international development was recognized as a matter of principle, civil society organizations felt that their status as participants and stakeholders on the international scene should be legitimized accordingly. The desire of CS organizations to be given post-WSIS tasks and responsibilities was thus unquestionably linked to their quest for political recognition.

Civil society organizations did not just criticize the lack of openness and transparency of the mechanisms proposed at the WSIS. Their perspectives offered a noticeably different vision of the roles and responsibilities of different actors and the structure and function of proposed implementation and follow-up mechanisms. CS insisted that whatever mechanisms were put in place should allow the WSIS to have a true impact on the national, regional, and international levels.

The diverging views between civil society and many government delegations were substantial and essentially concerned the following elements:

- the extent to which implementation and follow-up mechanisms adopted in Tunis should take the form of binding mandates for international organizations and governments;
- the extent to which the principles of multi-stakeholder participation, transparency, and openness would be reflected in the implementation and follow-up mechanisms;
- the roles and responsibilities of various stakeholders;
- the question of whether or not new institutions should be developed to ensure the implementation and follow-up of WSIS resolutions or if these responsibilities could/should be restricted to existing bodies;
- the very definition of implementation and follow-up mechanisms that was to be adopted at the WSIS.

Negotiating Follow-up Measures

The first political discussion on these issues was held during the second PrepCom as governments turned their attention to negotiation of what later became the Tunis Commitment and the operational plan, which later became the Tunis Agenda.

The Group of Friends of the Chair (GFC) took the initiative of submitting a draft text as a basis for negotiations.[6] The framework proposed by the GFC included a number of stocktaking and follow-up measures. It also promoted a multi-stakeholder solution to implementing the results of the WSIS. Amongst a series of initiatives, the GFC called for the formation of a separate team of stakeholders to work on each of the action lines included in the Geneva and, eventually, Tunis plans of action. In turn, the Secretary-General of the UN would be requested to nominate existing UN agencies or specialty

---

6 GFC, *Operational part of the final document / Tunis Agenda for Action / Tunis Plan of implementation.* (January 11, 2005). http://www.itu.int/wsis/gfc/docs/4/operational-part11jan-pm.html

organizations to either "moderate" or "coordinate" these multi-stakeholder teams. Selections of UN moderating/coordinating organizations would be made based on determinations about which organizational mandates and experiences were closest to the main theme of the action line. The UN organizations placed in these roles would then be expected to prepare reports on progress made and submit them to an (as yet) undefined body charged with overall coordination of the WSIS implementation efforts. Whatever organization occupied this overarching coordination role would in turn be responsible for regularly reporting back to the UNGA on progress made. Organizations mentioned as candidates to fill the overall coordination role included existing UN structures (ex. ITU, WSIS executive secretariat, UN Department of Economic and Social Affairs [UNDESA]) or a dedicated structure of some forum to be created expressly for this purpose.

The GFC text was the basis of subsequent negotiations on implementation and follow-up. Civil society, in comments overlapping with those of the CCBI, objected to the extent to which the proposed structures subjected the multi-stakeholder teams first to their UN coordinators and then again to some of the strictly intergovernmental bodies that were being proposed for an overarching coordination role. Instead, civil society and the private sector both proposed that the multi-stakeholder teams nominate their own coordinators from amongst the participating organizations and that the organization or structure put in the overarching coordination role reflect the multi-stakeholder principles of WSIS. A joint civil society / private sector statement was devised and issued demanding more meaningful multi-stakeholder participation in implementation and follow-up activities and in the deliberations through which they were being established.[7] These views were not shared by certain government delegations who sought explicit recognition of the jurisdiction of intergovernmental organizations, the ITU in particular.[8] As no consensus was reached at PrepCom II, the work of the GFC continued and, after several rounds of revisions, a much different document was presented as the basis for negotiations at PrepCom III.

---

7   Civil Society Plenary and the Coordination Committee of Business Interlocutors, *Joint statement on behalf of Civil Society Plenary and the Coordination Committee of Business Interlocutors on Implementation and Follow-up mechanisms for the WSIS post-Tunis.* (February 25, 2005). http://www.itu.int/wsis/docs2/pc2/Subcommittee/Jointcs-ccbi.html

8   Secretary-General of the International Telecommunication Union (ITU) and Director-General of the United Nations Educational, Scientific, and Cultural Organisation (UNESCO), *Possible Implementation Mechanism at the International Level* (WSIS-II/PC-2/DT-2(Rev.3)). (May 31, 2005). http://www.itu.int/wsis/docs2/pc2/working/dt3rev2.doc

For its fifth, sixth, and seventh meetings, the GFC compiled stakeholder contributions, including papers prepared by the preparatory committee president Janis Karklins. The president played an active role in the process, inviting government delegates to discuss concrete proposals and orienting the discussions towards issues of implementation and follow-up. The contribution that he presented at the seventh meeting was strongly criticized by several civil society actors, who objected to their lack of influence on the process and believed that certain modifications that he proposed to the operational documents diluted plans for implementation and follow-up mechanisms and made them vague and difficult to pin down.

> The draft "operational part" of the Tunis Summit declaration that had been produced by the "Group of the Friends of the Chair" (GFC) at PrepCom-2 in February was not great, but it was reasonable. It especially contained language that would ensure the continuation and expansion of the inclusive process that had distinguished the WSIS from previous Summits. It also would have included a serious implementation mechanism to ensure the Summit did not just produce tons of documents and travel expenses, but actually has an impact on the reality of the global information society. The latest draft of the "operational part," released by PrepCom president Janis Karklins on 16 August, is a serious setback compared to half a year ago. It will be discussed at the upcoming meeting of the WSIS GFC in Geneva on 5-7 September. Civil Society groups have no access to these meetings and are only allowed to comment at the open consultations on 6 September.[9]

In terms of its treatment of implementation, the new draft, distributed in August 2005, barely resembled its predecessor. Concrete mandates for multi-stakeholder teams were replaced by vague acknowledgment that: "Coordination of multi-stakeholder implementation activities would allow information exchange and avoidance of duplication of activities" (14d); "The establishment of multi-stakeholder partnerships...should be supported and encouraged" (at 14e); and "Each UN agency..., could facilitate activities among different stakeholders, including civil society and the private sector, to help national governments in their implementation efforts" (14b). The process of facilitating acitivies would have to be accomplished efficiently however, as the UN agencies were not permitted additions to existing budgets. Creation of a new organizational mandate to coordinate WSIS implementation activities was replaced with a request to the "Secretary General of the United Nations to submit a report on implementation activities of the WSIS decisions within the

---

9   Ralf Bendrath, *After Tunis: A Summit Without Implementation and Civil Society? Committment to Implementation and Multi-Stakeholder Approach Dropped from Documents, Civil Society Groups Voice Protest*. (September 2, 2005). http://www.worldSummit2003. de/en/web/781.htm

UN family as part of the annual report to the ECOSOC and/or UNGA."[10] The previously systematic, binding, official, and rigorous mechanisms of follow-up and implementation had been replaced by procedures that hinged on the goodwill of interested parties. In the demotion from "will" and "must" to "could" and "might," civil society's stake in follow-up and implementation rapidly shifted from wanting to maximize its own role in concrete processes to pushing for some form of binding commitment to solving problems underlined during the WSIS, as the political will of governments to do so was clearly plummeting.

The August 2005 draft provoked consternation from certain influential sectors of civil society. Recalling the decision taken by CS during Phase I to opt out of the official documents and produce an alternative declaration, Bertrand de La Chapelle suggested that the appropriate response from CS in this case might also be "a very simple 'exit' strategy/threat: if the final document appears too weak in terms of commitment...it will be very easy to issue a public statement exposing it and denouncing the incredible waste of time and money of the last two years."[11]

However, a surprise greeted participants at PrepCom III, as the Russian delegation objected to the adoption of the GFC text as the official negotiating text. This was particularly unexpected given that Russia had been an active member of the GFC. As a result, the GFC text became, rather than the exclusive negotiation document, one of a series of texts, alongside the PrepCom II version and others transmitted to PrepCom III, Subcommittee B for consideration.[12] Given these suddenly broadened parameters, civil society's focus shifted back to the negotiations. A CS "Working Group on Subcommittee B" was hastily formed in CS Content and Themes with Bertrand de La Chapelle as its coordinator and most visible spokesperson (or, as this role came to be described in the WSIS CS lexicon: "focal point"). This working group was mandated to:

> monitor discussions in Subcommittee B and report for actors not present in Geneva
>
> prepare statements and interventions and select speakers to present them in Subcommittee B

---

10 President of the PrepCom of the Tunis Phase, *Report on the Work of the GFC during the Inter-Sessional Period* (WSIS-II/PC-3/DOC/6-E). (September 8, 2005). http://www.itu.int/wsis/docs2/pc3/off6.doc

11 Bertrand de La Chapelle, *Revised GFC Draft–Strategic Aspects.* [WSIS CS-Plenary]. (September 17, 2005).

12 See Bertrand de La Chapelle, *GFC Document on Chapter One and Four Not Adopted as Basis for Sub Committee B.* [WSIS CS-Plenary]. (September 19, 2005).

advocate CS positions to other stakeholders, including governments

help define elements that could form part of future Civil Society declarations if the negotiations do not evolve in a satisfactory way

report every day on its activities, as appropriate in the Plenary and the Content and Themes meeting

liaise with the Chair of Subcommittee B on procedural matters and with the Internet Governance Caucus on potentially complementary issues.[13]

The purpose of the new CS working group was defined as to "help establish an efficient and flexible follow-up framework for WSIS, that guarantees the full and effective involvement of all stakeholders and particularly civil society at local, regional and international levels."

Negotiations at PrepCom III concluded without a consensus on implementation and follow-up among government delegations. Thus, as was the case with Subcommittee A on Internet governance, Subcommittee B of PrepCom III was reconvened in Tunis immediately before the summit. Negotiations there resulted in the adoption of the Tunis Agenda for the Information Society wherein paragraphs 83–122 deal with follow-up and implementation.

In the end, the outcome reflected something of a compromise between the concrete measures described in great detail in the PrepCom II documents and the minimal set of conditionally supported initiatives subsequently proposed by the August 2005 document.

The Tunis Agenda for the Information Society outlined a set of initiatives and procedures for three different levels of intervention: national, regional, and international.[14] Governments were fundamentally responsible for ensuring implementation at the national and regional levels and were mandated with the option of taking a series of measures defined in paragraphs 100 and 101. Members of civil society criticized the emphasis that was placed on allocating these responsibilities exclusively to state actors. First, the non-binding nature of the initiatives explicitly underlined the fact that states were only lukewarm in their commitment to following up on the WSIS and gave them the opportunity to contribute selectively to the implementation of the WSIS agreements. Second, civil society deplored being thus placed under the whim of government initiatives, especially after having repeatedly insisted on the development of a multi-stakeholder implementation and follow-up process. Civil society organizations were stripped of any power to take initiative as a

---

13   Bertrand de La Chapelle, *Invitation to Join the Working Group on Sub-Committee B*. [WSIS CS-Plenary]. (September 20, 2005).
14   These initiatives were defined in paragraphs 100 through 102.

result of the Tunis Agenda. Paragraph 100 of the Tunis Agenda for the Information Society encourages governments "with the participation of all stakeholders and bearing in mind the importance of an enabling environment, to set up a national implementation mechanism" centred around national e-strategies, efforts to "mainstream" ICTs, drawing on bi-lateral trade agreements where required, and factoring ICTD issues into assessment and reporting processes.

Implementation and follow-up measures took on three dimensions at the international level: an intergovernmental dimension, an inter-agency dimension, and a multi-stakeholder dimension. Civil society deemed this approach inadequate and reaffirmed the importance of applying the multi-stakeholder principle at all levels. The idea of multi-stakeholder facilitation teams working on action lines coordinated by UN agencies and departments was revived but made entirely optional and framed only as suggestion.

In accordance with paragraphs 102 and 103 of the Tunis Agenda, the implementation and follow-up measures were based on the action lines outlined in the official WSIS policies adopted by the Geneva Plan of Action. These action lines were the following:

- the role of public governance authorities and all stakeholders in the promotion of ICTs for development;
- information and communication infrastructure;
- access to information and knowledge;
- capacity building;
- building confidence and security in the use of ICTs;
- enabling environment;
- ICT applications:
    - e-government;
    - e-business;
    - e-learning;
    - e-health;
    - e-employment;
    - e-environment;
    - e-agriculture;
    - e-science;
- cultural diversity and identity, linguistic diversity and local content;
- media;
- ethical dimensions of the Information Society;
- international and regional cooperation.

The UN Secretary-General was asked to create a new group "with the mandate to facilitate the implementation of WSIS outcomes" (103). This would become the UN Group on the Information Society (UNGIS). The Tunis Agenda also requested that the United Nations General Assembly "proceed with analyzing the implementation of the conclusions of the WSIS in 2015" thus guaranteeing a space in which the UN could eventually be held to some measure of accountability for the results of the WSIS.

In other words, on implementation, expressions of interest in and support for follow-up were stronger than in the August 2005 version, but mandates were far more tepid and conditional upon volunteerism than they had been in the PrepCom II text. The stocktaking exercise undertaken in August 2004 and originally seen as an informal component of the larger discussion of implementation and follow-up held delegations' interest and gained momentum over the course of Phase II. It would become a lasting feature of the WSIS, as reports would be produced annually in order to measure the changing trends in the issues discussed at the summit and provide updates. Eventually, a permanent database was created and made available to WSIS participants and a "Golden Book"[15] outlining their commitments and activities was launched in Tunis. Thus, the language on monitoring and stocktaking exercises included in the Tunis phase official documents goes into considerable detail and the palpable enthusiasm evident for monitoring trends related to information society issues on an ongoing basis has to be seen as something of an unexpected positive outcome. In sum, the official texts did not turn out to be as catastrophic as civil society initially feared they could be. The WSIS demonstrated a genuine, albeit limited, desire to implement the decisions adopted in Geneva and Tunis.

WSIS Implementation and Follow-up: An Overview

The implementation and follow-up measures set up at the WSIS exist on the ground as a complex and multi-faceted set of occasionally intersecting, but often distinct initiatives. Here, we present a synthesis of the major initiatives that were developed within the framework of the WSIS.

---

15 See WSIS Executive Secretariat and ITU, *Golden Book Portal*. http://www.itu.int/wsis/goldenbook/index.html

160 • DIGITAL SOLIDARITIES •

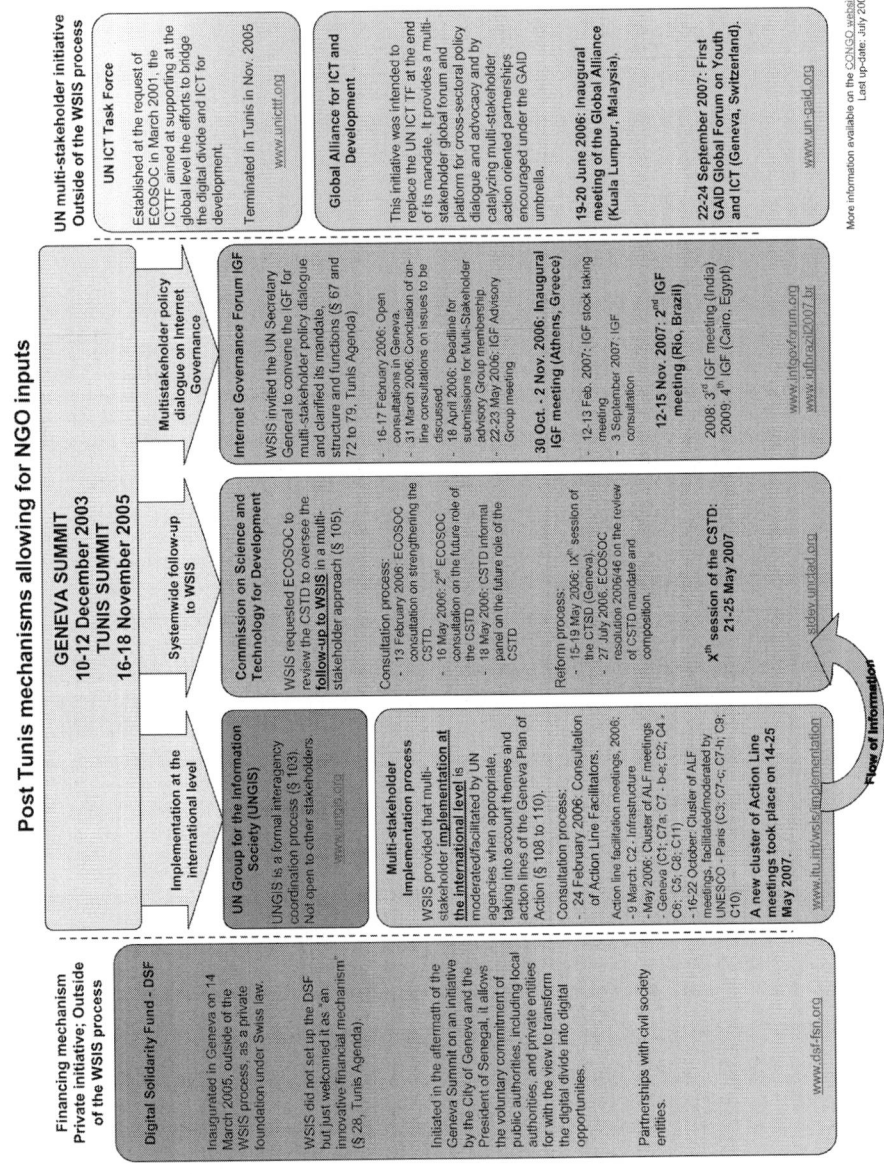

Figure 3: Post Tunis Mechanisms Allowing for NGO Inputs[16]

---

16 Produced by the Conference of Non-Governmental Organizations in Consultative Relationship with the United Nations (CONGO). http://www.ngocongo.org/congo/files/chart_on_post_wsis.pdf

United Nations Economic and Social Council (ECOSOC)

The WSIS implementation and follow-up effort largely hinges on the UN Economic and Social Council's legitimacy and authority. By mandating other organizations to act, the ECOSOC can adopt specific goals, take appropriate measures to improve the implementation and follow-up of WSIS decisions, and correct any shortcomings in established processes that may have been identified.

The ECOSOC has ultimate responsibility over the implementation and follow-up of WSIS decisions. The Council approach to its WSIS follow-up and implementation responsibilities is multi-faceted. First and above all, the Council draws on the Commission on Science and Technology for Development (CSTD) to assist it. This relationship was established not only in the WSIS documents (discussed below) but by virtue of the ratificaton of two UN resolutions calling for UN agencies to play the roles of coordinators/facilitators of the thematic WSIS action lines, and regional commissions of the United Nations and other relevant bodies to periodically submit reports to the CSTD on the implementation of WSIS decisions.[17] Thus, the CSTD is charged with evaluating the progress of implementation and follow-up and with proposing initiatives aimed at improving their efficiency to the Council.

By compiling and analyzing all of the documentation that is transmitted to it by the main organizations in charge of implementing the decisions of the WSIS, the CSTD produces general evaluations of the implementation measures annually. The ECOSOC then adopts the necessary dispositions for improving their efficiency based on these reports and decides which measures should be taken in order to achieve the goals of the WSIS in accordance with the Geneva Plan of Action and the Tunis Agenda.

Commission on Science and Technology for Development (CSTD)

The Tunis Agenda entrusted the ECOSOC with reviewing the mandate, agenda, and composition of the CSTD in order to turn it into the main institution responsible for coordinating the implementation and follow-up meas-

---

17 See The United Nations General Assembly, *Resolution A/RES/57/270 B: Integrated and Coordinated Implementation of and Follow-up to the Outcomes of the Major United Nations Conferences and Summits in the Economic and Social Fields.* (July 3, 2003). http://www.unctad.org/en/docs/ares57270b_en.pdf See also the General Assembly, Resolution 60/252: *World Summit on the Information Society.* (April 27, 2006). http://www.itu.int/wsis/docs/background/resolutions/60-252.pdf

ures of the WSIS at the international level. The role of the commission in this respect is the following:

- compiling the reports of different partners, more specifically those of international organizations in charge of implementing the different policies of the Geneva Plan of Action, and producing periodic reports on the implementation and follow-up measures of the decisions adopted at the WSIS;
- drafting proposals for resolutions to be adopted by the ECOSOC based on these reports, demanding that it mandate different international organizations to undertake appropriate actions to (1) improve the implementation and follow-up mechanisms already set up, (2) correct any shortcomings in the implementation process, and (3) reinforce the coordination of the actions taken on all levels.

The ECOSOC therefore has to evaluate the reports and recommendations proposed by CSTD and, if necessary, adopt the resolutions needed to improve the implementation process set up at the regional and international levels. The resolutions then serve to orient the activities of different regional and international organizations toward the implementation and follow-up of the WSIS.

The composition of an international moderators/facilitators group for the different WSIS Action Lines

In accordance with the Tunis Agenda, the responsibility for implementing the 11 action lines put forward by the WSIS was divided among various international organizations placed in charge of organizing meetings, events, and multi-stakeholder gatherings related to the themes they were assigned. These organizations periodically draft reports on the progress that has been made on the implementation measures within their respective themes. The reports serve as the basis for the work of the CSTD.

The UN Group on the Information Society (UNGIS)

The Tunis Agenda requested that the Secretary-General of the United Nations establish a UN Group on the Information Society (UNGIS) in consultation with members of the UN Chief Executives Board for Coordination (CEB). The Group was put in charge of incorporating the WSIS agenda into the activities and programs of various UN organizations that were members of the CEB.

UNGIS was officially established by then United Nations Secretary-General Kofi Annan. Its creation was ratified in April 2006, five months after the end of the WSIS[18] and was endorsed in April 2006 by the CEB as a new inter-agency initiative for coordinating the implementation measures of the WSIS decisions that had been adopted by UN agencies. The UNGIS's tasks, among other things, are to coordinate the implementation efforts of different UN agencies, prevent the duplication of initiatives, settle public policy issues resulting from the implementation of WSIS decisions within the UN system, and contribute to informing the public on the challenges facing the information society.

The UNGIS groups together organizational members of the CEB whose mandates somehow reflect on or interact with information society issues. It holds annual meetings. Every year, the UNGIS elaborates a limited number of themes relating to the information society and the WSIS and focuses on the policies and initiatives set up by a few selected countries in order to develop a toolbox for orienting ICTs in the service of international development.[19]

Implementation strategies at the national, regional and international levels

Various initiatives were presented in the Tunis Agenda in order to support the implementation of the decisions adopted at the WSIS at the national, regional, and international levels.

At the national level, the Tunis Agenda encouraged governments to establish national implementation mechanisms that would help achieve the UN Millennium Development Goals and establish joint initiatives with relevant partners. At the regional level, it called for regional intergovernmental organizations to undertake implementation activities in conjunction with other partners, exchange information and best practices, and discuss public policy issues related to putting ICTs in the service of development. Governments also requested regional commissions of the UN to contribute to the organization of follow-up activities. The multi-stakeholder principle was deemed "essential" to the organization of these activities. Finally, at the international level, the Tunis Agenda called for a strategy based on the policy themes outlined in the official documents of the WSIS to be implemented by different

---

18 The following organizations are members of the UNGIS: ECLAC; FAO; IAEA; ILO; ITU; OECD; UNCTAD; UNDESA; UNDP; UNECA; UNECE; UNESCO; UNESCWA; UN-HABITAT; UNHCR; UNICEF; UNIDO; UNITAR; UNODC; UNRWA; UNWTO; UPU; WB; WFP; WHO; WIPO; WMO; WTO. See http://www.ungis.org.

19 For more details on the UNGIS as well as a full list of the UN departments and agencies that form its membership and contribute to its work, see its website http://www.ungis.org/

UN agencies according to their respective mandates and resources. All of this was to be done following the multi-stakeholder principle.

Stocktaking (ITU)

Launched in October 2004, the "stocktaking" procedure is a statistical compilation of the different initiatives adopted by the various parties participating in the WSIS regarding the different themes prioritized by the summit.

The ITU annually publishes information about the various initiatives for reducing the digital divide developed by all stakeholders. Participation in this activity is voluntary. Stocktaking involves the allocation of different responsibilities, which link Phase I decisions to different governmental, institutional, private, and non-governmental partners. Additionally, it informs the different partners about who is doing what in regards to which domains/fields of specific activities/problems, and how efficiently. The ITU produces an annual report on different stocktaking activities and presents data related to the main themes identified by the Geneva Plan of Action.[20]

Statistics and Indicators

During the WSIS it was seen to be fundamental for implementation and follow-up work to be able to draw on high-quality reliable statistical indicators that would be capable of identifying up-to-date problems related to the digital divide, including its extent, dimensions, and principal causes, its tendencies and developments, and the problems related to integrating new information and communication technologies into development issues.

The process of monitoring the development of the information society as it was established by the Tunis Agenda rested upon the development, reinforcement, and improvement of the data, indicators, and indexes that measured the main factors concerning issues of connectivity, access, penetration, use, costs, and availability of ICTs, particularly with respect to developing countries. The Tunis Agenda called for periodic evaluation of the problems identified at the WSIS (paragraphs 112 and 113), outlined a series of criteria for indicators to be used in these evaluations (113-117), and encouraged the international community to combine to their statistical data capacities (paragraph 118).

This approach has led to the drafting and distribution of annual reports and the creation of statistical methods that reflect the development of the in-

---

20 For details on the process and copies of the various annual reports, see the WSIS Stocktaking database at http://www.itu.int/wsis/stocktaking/index.html

formation society. The reports track the progress that has been made in the fight to reduce the digital divide as well as the steps being taken to implement the results of the WSIS.[21] The WSIS also resulted in the establishment of the Partnership on Measuring ICTs for Development, a multi-stakeholder body meant to lead to greater availability and quality of indicators linking NICTs to development. This initiative grouped several international organizations together around the production, sharing, and improvement of data and indicators related to issues in international development, technical development, and integration. The WSIS endorsed two statistical indexes based on indicators defined in the Partnership on Measuring ICTs for Development initiative. The indicators, called the ICT Opportunity Index and the Digital Opportunity Index (paragraphs 114 and 115), are defined and presented in the WSIS annual reports. They evaluate and measure the efficiency of the implementation measures set up by the WSIS.[22]

Implementing Lessons Learned

One of the main weaknesses of the WSIS was its inability to ensure that decisions taken by governments would be implemented concretely. Despite this, however, initiatives undertaken by WSIS participants (from civil society and developing countries in particular) since the closing of the WSIS attest to the existence of a desire to implement in practice the principles discussed at WSIS in order to contribute to bridging the digital divide.[23]

Action on regional initiatives that would ensure the implementation of the WSIS also depends heavily on the will of states to mandate the regional

---

21 See ITU and UNCTAD, the *World Information Society Report: Beyond WSIS*. (June 2007). http://www.itu.int/osg/spu/publications/worldinformationsociety/2007/WISR07_full-free.pdf

22 The ICT Opportunity Index is the result of the merger of the ITU's Digital Access Index (DAI) and Orbicom's Monitoring the Digital Divide/Infostate conceptual framework. The ICT Opportunity Index is an inclusive tool that measures economies' ICT networks, skills, and use. http://www.itu.int/ITU-D/ict/publications/ict-oi/2007/index.html. The Digital Opportunity Index is an e-index based on internationally-agreed ICT indicators. This makes it a valuable tool for benchmarking the most important indicators for measuring the Information Society. The DOI is a standard tool that governments, operators, development agencies, researchers and others can use to measure the digital divide and compare ICT performance within and across countries. http://www.itu.int/ITU-D/ict/doi/index.html

23 The details of these initiatives can be found in report A/63/72-E/2008/48. See United Nations Economic and Social Council, *Progress Made in the Implementation of and Follow-up to the Outcomes of the World Summit on the Information Society at the Regional and International Levels* (A/63/72-E/2008/48). (April 2008). http://www.unctad.org/en/docs/a63d72_en.pdf

organizations to which they belong. However, there was an inevitable loss of momentum following the closing of the official work of the WSIS and the will of states to act voluntarily may never materialize. Nevertheless, several economic cooperation and development institutions in the Americas, Asia, Africa, and Europe have embarked on regional initiatives that have contributed to the implementation of the Geneva and Tunis principles.[24]

At the international level, time, and the WSIS review scheduled for 2015, will tell whether the administrative mechanisms undertaken by various UN departments and agencies are leading to concrete results or merely shuffling a largely bureaucratic deck.

Overall, the lessons learned by civil society from the debate over WSIS follow-up and implementation should not soon be forgotten and should guide future policy intervention in this and other venues. Most fundamentally these lessons are that the substance of issues debated can be easily undermined in the details about concrete follow-up and that, absent firm commitments of political will to facilitate meaningful implementation from governments, civil society can and does bear the brunt of burden for making sure that policies actually connect with people. For this reason, on the issue of implementation—perhaps above any other—civil society speaks with unquestioned moral authority when demanding government action.[25]

---

24   Ibid.
25   For a discussion of how this problem plays out at the national level, see also Marc Raboy and Jeremy Shtern, *Media Divides: Communication Rights and the Right to Communicate in Canada*. Vancouver: UBC Press, 2010.

PART THREE
Civil Society and Global Communication Governance Beyond WSIS

The WSIS ought to be considered both as an experiment in global communication governance and a political marker. As a multi-stakeholder experience, the event tested the effectiveness and feasibility of integrating non-government actors into an intergovernmental political negotiation process. As a political marker, the WSIS set a new level—theoretically at least—for the participation of NGOs in subsequent political negotiations.

The political and institutional legacy of the WSIS will thus be largely judged by the role the summit played in the democratization of global communication governance going forward. Understanding this legacy involves reflection on this emergent governance model and asking questions about how CS participation in the global governance of communication has evolved post-WSIS.

The last three chapters of this book will address WSIS CS activities in other international organizations, examine the Internet Governance Forum process that was created by the WSIS, and provide an in-depth critical assessment of the role of civil society within global communication governance nearly five years following the conclusion of the WSIS.

• CHAPTER SEVEN •

Civil Society, Internet Governance and the IGF

As we described at the outset of this volume, the WSIS was not just a summit held in two phases, it was—in particular where civil society was concerned—a two-phased process.

During Phase I, the Communication Rights in the Information Society (CRIS) campaign played what the 2005 companion volume to this book described as a "catalyzing role"[1] in helping to organize civil society. CRIS's influence was manifest as of the first WSIS PrepCom in July 2002, which was, as we described it,

> a time of great confusion and frustration for civil society's actors, as they could only take note that the doors of a so-called multi-stakeholder Summit were being closed one after the other in front of them. CRIS took on an unofficial function of facilitating dialogue, mobilization and consensus building among the different organizations. In fact, the CRIS presence at PrepCom 1 was so strong that it attracted criticism from other actors who feared a subordination of civil society participation to CRIS's own agenda.

Mueller and his colleagues are more specific in their analysis, pointing out that it was "CRIS principles" that, for the most part, occupied coordinating and leadership roles of the important CS structures over the course of the first phase of the WSIS.[2]

The turnover in civil society participation had already begun late in the first phase of the WSIS, as certain individuals stepped back or dropped out and new interested civil society participants emerged as the process unfolded. This shift has been systematically documented by Cammaerts and Carpentier. Their analysis suggests, in particular, that the participation of civil society or-

---

1 See Marc Raboy and Normand Landry, *Civil Society, Communication and Global Governance: Issues from the World Summit on the Information Society.* New York: Peter Lang, 2005, at p. 42.
2 Milton Mueller, Brenden N. Kuerbris, and Christiane Pagé, "Democratizing global communication? Global civil society and the campaign for Communication Rights in the Information Society." *International Journal of Communication* 1, 2007 (at 282).

ganizations increased 41% between PrepCom III and the Geneva Summit itself. In other words, literally hundreds of civil society organizations that had not participated in the preparatory process were drawn to the summit.[3] While systematic analysis of exactly how many of these latecomers would then go on to meaningfully participate in the second phase of the WSIS is not readily available, it is clear that the constitution of WSIS civil society changed greatly between Phases I and II, because of the influx of new participants drawn to the process as a result of attending the Geneva Summit, but for a number of other reasons as well. There was natural process fatigue amongst individuals who had participated intensely in the WSIS for the better part of two years, certain organizations lacked the resources to continue sponsoring their members or employees to participate or simply had other projects demanding attention, and the WSIS agenda itself narrowed and focused on certain issue areas—Internet governance above all—that might not have seemed as relevant to the interests and mandates of certain CS actors and organizations who had been centrally involved in Phase I.

The extent of the changing of the CS guard that occurred between the end of Phase I and the early stages of Phase II can be observed empirically through a cursory examination of the individuals occupying key CS leadership positions including coordination of caucuses, the content and themes group and the CS plenary. It can also be observed through comparing who posted messages with the greatest frequency on the key CS email lists in Phase I and Phase II. More systematically, in presenting their social network analysis of WSIS civil society participation, Mueller et al. address the emergence of a "second generation" of WSIS CS.

Mueller et al.'s social network analysis suggests that, by Phase II, the Association for Progressive Communications (APC) had eclipsed the CRIS campaign as the organizational hub for civil society participation in the WSIS. The fact that APC was both a founding member and key figure in the CRIS campaign should not be discounted.[4] But Mueller et al.'s analysis of the level of individual participation is more revealing. Of the 15 individual WSIS participants judged during Phase II to have been most central to the network of WSIS civil society participants, only 2 identified themselves as members of the CRIS campaign and listed communication rights as one of their primary issue

---

3   See Bart Cammaerts and Nico Carpentier, *The Unbearable Lightness of Full Participation in a Global Context: WSIS and Civil Society Participation*. 8. Media@lse, London, UK, 2005. http://eprints.lse.ac.uk/4037/

4   For example, APC organizer Karen Banks, one of the most active civil society participants in the WSIS (and a CS member of the WGIG), was also a member of the CRIS campaign steering committee.

areas of interest. In contrast, 8 of the 15 most central individuals listed Internet governance as a primary issue area of interest, of which 4 listed Internet governance as their only issue area of interest at the WSIS.[5] Mueller et al. attribute this to the emergence of Internet governance as a preoccupation for the agenda of the second phase of the WSIS. They further suggest that "with the exception of APC, no major CRIS-affiliated actors had been involved in Internet governance or had knowledge of the institutions and issues" but also conclude that this shift in focus had the effect of "br[inging] to the fore civil society actors from ICANN-related issue networks and the WSIS-CS Internet Governance Caucus."[6] To these countervailing trends they attribute the presence of what they call a "particularly evident gap" between CRIS-affiliated and Internet governance-affiliated actors and organizations.[7]

In our previous book on the first phase of the WSIS, we framed the CRIS campaign as in many ways emblematic of CS participation in WSIS Phase I. In a similar vein, the Internet Governance Caucus can be seen to have defined and dominated civil society participation in the second phase of the WSIS. These two loose frameworks represent very different approaches to coordinating CS participation in global governance. The CRIS campaign used the normative banner of communication rights and the opportunity of the WSIS to rally a previously disparate array of CS actors working in distinct fields and on different issues but sharing a more or less common commitment to social justice[8]. The organizing principle of the IGC, on the other hand, was a shared interest in a specific policy field and set of issues, and its membership reflected wildly divergent and often competing normative agendas related to the goals for Internet governance.

As it emerged as the policy sphere in which civil society participation in Phase II of the WSIS was arguably the most developed, we will now examine how civil society structures for participating in the global Internet governance debate have evolved since the WSIS.

---

5   Milton Mueller, Brenden N. Kuerbris, and Christiane Pagé, "Democratizing global communication? Global civil society and the campaign for Communication Rights in the Information Society." *International Journal of Communication* 1, 2007 (see chart at 287).
6   Ibid (at 289).
7   Ibid (at 290).
8   See, for example, Sean Ó Siochrú, "Implementing Communication Rights." In Marc Raboy and Jeremy Shtern, *Media Divides: Communication Rights and the Right to Communicate in Canada*. Vancouver: UBC Press, 2010.

Addressing the Democratic Deficit in CS? The Internet Governance Caucus Charter Drafting/Adoption Process

Internet governance is one area where the participation structures created by civil society during the WSIS endured beyond its conclusion. The IGC has emerged as a focal point for CS participation in the debate over global Internet governance. Like the debate itself, the CS Internet governance caucus has evolved and matured since Tunis. Efforts have been made to apply the lessons learned over the course of the WSIS and, most fundamentally, to take steps to address the challenges of providing some measure of legitimacy, transparency and democracy to civil society participation in global governance. Such questions have dogged civil society, the IGC in particular, since the WSIS. Consideration of what the IGC reforms have accomplished and failed to achieve is revealing about the more general prospects for efforts to improve the organizational structures of CS.

At Phase II, PrepCom II in Geneva, on February 23, 2005, one of the founders of the CS Internet Governance Caucus (IGC) took the floor at the civil society Content and Themes meeting to criticize a statement that had been made by the IGC. The content of this complaint was that the positions of developing country governments on the issues related to Internet governance were being spurned by civil society in favour of views that were more supportive of the private sector. Its substance was, however, more salacious, controversial and revealing.

The intervener, a former IG caucus coordinator, charged that a group of individuals, each based out of OECD countries and heavily involved in ICANN processes, had formed a clique of "opinion leaders" in and around the caucus coordinator positions and were setting a narrow and problematic agenda for the caucus's interventions and, by extension, for civil society's position on the issue of Internet governance. The complaint was about the transparency of the processes through which the IGC's PrepCom II statement had been drafted and consented to by WSIS civil society and the legitimacy of the structures which enabled certain civil society delegates to assume leadership roles.[9]

Like many of the other CS caucuses, the IGC had been an ad hoc and largely informal structure devised during the early days of the first phase of the summit in the effort to set aside institutional and discursive space for coordinating the deliberation and intervention of CS actors on Internet governance.

---

9   See Y.J. Park, *YJ's Objection and the CS-PS Statement*. [WSIS CS-Plenary]. (February 24, 2005).

There were no formal rules established regarding membership or the selection of its leadership, nor was there a clear set of guidelines in place for determining the process by which "the" caucus position on any given issue was arrived at. Anyone was free to subscribe to the IGC listserv or attend its meetings. Caucus coordinators were established by rough consensus, which was also used to achieve caucus approval for any interventions to be made on its behalf. These working modalities were common and largely unproblematic across a large number of the WSIS CS caucuses.

However, by Phase II, PrepCom II, the IGC was in many ways distinct from the other CS caucuses. The broadly framed and highly controversial issue of Internet governance was now a central focus of the WSIS. As a result, in order to measure consensus, the IGC was obligated to consider not only the 170-200 people who were subscribed to its email listserve at various times but also the inputs of other WSIS CS caucuses (Human Rights, Gender, Africa, etc.) which were increasingly finding their own issue areas subsumed by the IG negotiations.

As the attention of the WSIS increasingly focused on Internet governance, the CS Internet Governance Caucus emerged, over the course of the second phase of the summit, as not only a focal point for civil society input to the negotiations, but as a site of discussion and reflection on how its own largely ad hoc structures and working methods could be reformed to make civil society participation in global governance generally more open, transparent and democratic. With the conclusion of the WSIS process approaching, it was clear to both caucus members and other civil society actors that, by virtue of the impending establishment of the Internet Governance Forum, the IGC would play an important role in the continued involvement of the CS networks and structures that had been developed over the course of the WSIS—if they were to endure beyond Tunis.

It was also clear that existing WSIS civil society practices were not going to be sufficient. By Phase II, the CS Internet Governance Caucus convincingly met the four-part test required in order for new or existing CS caucuses and working groups to participate in civil society's crucial content and themes proceedings:

> Caucuses and Working Groups may participate as members of CTG [the Civil Society Content and Themes Group] if they can satisfy these conditions:
>
> 1. having a statement of intent,
> 2. a contact point and (partial) list of members,
> 3. at least one open meeting, preferably more, at the current WSIS-related event

4. a discussion list and open archive.[10]

But clearly more was required. The controversy around the Phase II, PrepCom II, IGC statement on Internet governance underlined a series of obstacles to effective participation in the WSIS created by the absence of more formal CS processes.

One of the aims of the CS caucus structure had been to allow for diverse viewpoints within CS to coalesce in more or less like-minded groups that would work and participate in parallel, thus embracing the diversity of civil society perspectives and assuring that the inevitable internal differences in opinion would enrich the WSIS debate rather than preoccupy civil society. Despite the fact that efforts were made to give a voice to the Human Rights Caucus, the Privacy Caucus, the Africa Caucus and any other interested CS groups in the Phase II, PrepCom III negotiations of an agreement on Internet governance, some sectors of civil society expressed frustration with the extent to which the perspective of CS writ large on the issue of Internet governance seemed to have become conflated with the view of the IGC. Not only was the ability of CS actors with views that diverged from those emanating from the IGC to use the caucus structure to participate in the Internet governance debate of Phase II somewhat limited, the advantages of participating in the IGC itself were undeniable. The IGC's issue focus seamlessly aligned with the political agenda of the WSIS by the later part of Phase II and, by virtue of their participation in the WGIG, a group of the caucus's core contributors had become visible, well connected and influential within the wider WSIS. The end result was that membership in the IGC increased over the course of the WSIS and, rather than a like-minded group that could be easily and informally administered, the caucus evolved into a large, diverse group reflecting views, backgrounds and agendas from all over civil society. That the efforts of the IGC proved to be fractious, controversial and perpetually subject to criticism suggests that the caucus structure simply may not have been scalable to the inevitable evolution that would be required as CS participation expanded, matured and focused.

On one hand there were issues of internal agenda setting and gate keeping. CS processes were coming to be seen, at least by some, as prone to capture and manipulation and it was felt that more transparent and inclusive practices would be required to ensure truly representative consultation on developing

---

10 From the Conference of Non-Governmental Organizations in Consultative Relationship with the United Nations (CONGO), *Civil Society Orientation Kit*. (November 2005). http://www.ngocongo.org/congo/files/wsis_oriention_kit.pdf. See Chapter 3 for discussion of this process.

positions. Furthermore, greater clarity was needed in relation to the sticky issue of defining which voices from within CS various interventions were claiming to represent. In a related vein, it was evident that opportunities had to be made available for the expression of dissenting views.

A separate set of concerns were voiced about effective and efficient participation. Some of those who felt that they had—in good faith—participated in open deliberations around the IGC position for Phase II, PrepCom II expressed frustration about the ease with which all of their work could be undermined and discredited by one (or a handful) of individuals willing to grandstand their discontent with the results that had been achieved. The coordinators of the caucus, in particular, were frustrated by the implication that they had overstepped non-existent boundaries.

There was also the irony of the general civil society thematic critique of the need for more open, transparent and legitimate procedures in global governance of the information society and the shortcomings of same in civil society's own internal processes. This point was frequently reinforced to civil society participants by government delegations including Norway and Cuba. The need to establish formalized written procedures concerning membership, leadership and the processes by which caucus positions were established was thus intensely debated in the aftermath of Phase II, PrepCom II.

The issue of establishing legitimacy, transparency, and democracy faced by the IGC was common, to varying degrees, to all WSIS civil society structures. However, in the general absence of such processes, the prevailing practices were largely justified by the need for ad hoc, pragmatic structures in order to get to the end of the WSIS. With the post-WSIS establishment of the IGF, it was clear that the IGC—or some similar structure—would be required, if not permanently, then at least on an ongoing basis. With the conclusion of the WSIS and the emergence of the IGF as a venue for continuing the debate over Internet governance—as well as emblematically for continuing the campaign for civil society participation in multi-stakeholder global governance of communication in general—the Internet Governance Caucus began a serious process of reflection on and reform and formalization of its processes in early 2006. The primary venue for this discussion was the IGC email list.

While there was broad consensus from both caucus supporters and critics alike that more formalized rules were required in order to develop more open, democratic, transparent and legitimate processes, there were—even amongst caucus core contributors—divergent views on what functions those processes should be supporting. Most interventions identified three possible roles for the caucus:

- provide an open forum for discussing Internet governance issues of importance to civil society, in particular issues on the agenda of the IGF;
- nominate speakers and organizing committee representatives etc. for the IGF on behalf of civil society;
- allow like-minded groups of individuals to develop common, consensus-based positions on Internet governance issues and collectively draft interventions for the IGF and other IG institutions.

Only the first of these roles was entirely unproblematic and the issue of the desirability and mechanics of establishing consensus-based caucus positions proved particularly divisive.

In March 2006, amidst the backdrop of an ongoing discussion of general caucus reform, the convening of the IGF's Multi-stakeholder Advisory Group (MAG) provided an opportunity for the IGC to make the first post-WSIS moves towards more formalized transparent processes. The immediate issue concerned the nomination of CS representatives to the MAG.

On the basis of listserv discussions, a core group of caucus members developed and refined a process for nominating civil society representatives to the MAG based on the Internet Engineering Task Force (IETF) "Publicly Verifiable Nominations Committee (NomCom) Random Selection" process.[11] The IGC set up its own version of the NomCom, meaning that a committee would be randomly formed for volunteers to evaluate those who were nominated and who nominated themselves to fill the 10 or so slots thought to be reserved for CS on the MAG. The NomCom was designed to ensure that nominees would be selected by merit and that geographical, gender and other UN-required balance would be achieved. However, in order to pre-empt charges of cliquish behavior and agenda setting, the committee that would evaluate candidates for nomination would be chosen randomly. The process contained the following steps:

- Caucus members were encouraged to volunteer to serve on the NomCom, the idea being that true random selection required a large number of volunteers. It was determined that the minimum number of NomCom volunteers required in order to randomly select the 5 eventual members of the committee was 30.
- Each volunteer was assigned a number based on the order that their offer to serve was received (i.e., the first person who offered was assigned 1 and the 30th person was assigned 30 etc.).

---

11 See the Network Working Group, *Publicly Verifiable Nominations Committee (NomCom) Random Selection*. (June 2004). http://www.ietf.org/rfc/rfc3797.txt

- A number set that would not be publicly known until after the volunteer period was concluded was identified at the onset of the NomCom (in this case the numbers selected in the April 1st, 2006 draws of the Irish national lottery, UK national lottery and US Powerball lottery were chosen as the "seed" number set).
- A complicated mathematical formula was performed that distilled these seed numbers into another set of numbers that corresponded to the list of volunteers. This group of people became the randomly selected nominating committee.
- Caucus members were invited to put forward nominations or to nominate themselves for the MAG. Anyone who was selected to serve on NomCom was not eligible to be considered, but volunteers for the NomCom random selection process who were not selected to serve on the NomCom were under no such constraints.
- The NomCom subjectively considered the relative merits of the list of 28 nominations and recommended a slate of 15 candidates for the civil society positions on the MAG.

Thus, the process became more transparent—in the sense that there were clear and established rules—and legitimate, in the sense that steps were taken to insulate the decisions made by the caucus from accusations of capture.

Yet there will still issues. The first was that, in spite of the role that random selection played in creating the nominating committee, the nominations themselves would still have to come from the selective choices of individuals based on some set of criteria. Defining the criteria would then play as much role in setting the agenda of the NomCom as establishing its members would.

Secondly, there was the relationship of the IGC to the other structures of WSIS civil society and to the larger category of broadly defined civil society actors who had not participated in the WSIS for whatever reason but who might, could or even should want to participate in the IGF.

Thus, it was agreed that more formalization of the IGC caucus structure was required going forward. The fundamental barriers to formalization of caucus processes, however, remained the question of how membership could be defined, and the related unresolved debate over whether or not the caucus should participate in collective deliberation and intervene as a single voice in IG debates.

Membership definition was not an entirely new question for WSIS civil society.[12] But the question of how membership in the IGC was to be defined

---

12 The WSIS Human Rights Caucus, for example, adopted membership criteria after the disruptions of its activities at Phase II, PrepCom I (see Chapters 2 and 3). The criteria

was particularly problematic given that the publicly available, open-subscription listserv of the caucus—where most of its work and discussion took place—had hundreds of subscribers, very few of whom were actually involved in caucus work or considered themselves to be members of the caucus. As Internet governance had emerged as an important issue over the course of the second phase of the WSIS, important actors from various IG-stakeholder groups—both involved and external to the WSIS—subscribed to the listserv in order to follow the debate and to benefit from the opinions of experts and specialists involved in the caucus. Because of this lurker population, the caucus had become an important interface between CS and other stakeholders and had developed a degree of brand power within WSIS circles. Any definition of membership based on subscriptions to the listserv would be unworkable. Any definition of caucus membership that restricted discussions to those who formally declared themselves to be members would exclude the lurker population and diminish the utility of the caucus. This distinction about whether membership would be defined by the entire listserv subscription list or by some subset of it was crucial to the legitimacy of whatever structures were to be created as there simply had to be some way of establishing the legitimacy of these new rules and frameworks.

After considerable discussion, it was decided that caucus membership could not be defined by subscriptions to the listserv. While 400 plus people might be subscribed to the list, it was determined that only a few dozen had ever been actively involved in caucus activities at a given time. Many of the subscribers were assumed to be government or private sector delegates to the WSIS/IGF who would not, under any circumstances, vote on or get involved in caucus reform discussions. Thus, to expect that any proposed caucus structures would need the affirmation of 50%+1 of all subscribers was felt to be simply impossible given that it was unlikely that that many subscribers would even vote. It was thus determined that the listserv would remain open as a discussion list, but that the adoption of new rules and a charter, the selection of coordinators and of other officers, participation in collective statement drafting and the approval of the contents of interventions to be made on behalf of the caucus would be functions restricted to members. For the purposes of charter ratification, membership was defined as those subscribers to the

---

specified that only groups were granted membership, precluding individuals from joining the Human Rights Caucus, and all members were required to agree to "the goal of protecting and promoting human rights standards in the WSIS process and in all countries of the world, not least the host country of the Summit." See Meryem Marzouki, *The Human Rights Caucus Stresses Major Advances Despite Attempts of Blockage.* (EN) WSIS. (July 3, 2004). http://www.iris.sgdg.org/actions/smsi/; hr-wsis/#3

listserv who presented themselves as members. Subsequently—if approved—the charter would define membership more clearly.

Through an at times arduous process, a charter was drafted and, on the basis of rough consensus, finalized. A voter's list was created based on those who presented themselves as caucus members. Ballots were distributed by email and the vote closed on October 2, 2006. Sixty-seven votes were cast, with 95% approving the charter. The text of the charter defines the vision of the caucus as well as its membership, working methods, decision-making processes, rules of conduct, organizational roles and the processes through which individuals should be elected to and recalled from such roles.[13] The IGC would emerge as the primary focal point for CS participation in the Internet Governance Forum (IGF).

Internet Governance Forum (IGF)

The most visible outcome of the WSIS process so far is the IGF, a new, innovative, semi-permanent organization that was created for addressing Internet governance and policy issues at the global level within the UN system. The IGF meets annually under the aegis of the Secretary-General of the United Nations. At time of writing, it has met four times: 2006 (Athens), 2007 (Rio de Janeiro), 2008 (Hyderabad, India) and 2009 in Sharm El Sheikh, Egypt. The mandate of the IGF calls for the Secretary-General of the United Nations to conduct a review and evaluate whether or not it will be extended beyond an initial five-year period.

The IGF has effectively functioned as a laboratory for experimentation with various approaches to multi-stakeholder global governance practice. The governance innovations associated with the IGF process typically receive enthusiastic backing from stakeholders—civil society and the private sector in particular—and versions of the IGF model of multi-stakeholder governance have been enthusiastically imported by various other international organizations. On certain issues, the IGF has proved to be an effective facilitator and coordinator of multi-stakeholder action. Yet, on balance its focus and impact is often restricted to process rather than substance. Regardless, evaluating the IGF's experience of creating a multi-stakeholder global governance forum from the ground up is essential to analysis of both the legacy of the WSIS and the trajectory of multi-stakeholder global governance of communication.

---

13   See The Internet Governance Caucus, *The Internet Governance Charter*. (Approved on October 2, 2006). http://www.igcaucus.org/charter

The Establishment and Anatomy of the First IGF

Consultations on the convening of the IGF were held in Geneva in February and then again in May of 2006. On 17 May 2006, the UN Secretary-General Kofi Annan established the Multi-stakeholder Advisory Group (MAG) to "assist him in convening the Internet Governance Forum."[14] The MAG would meet regularly and serve as a lightning rod for criticism in the lead-up to IGF 2006.

The composition of the MAG and its role in setting the IGF 2006 agenda were sources of intense controversy that was very clearly counter-productive to the goal of drawing on multi-stakeholderism in the effort to counteract the endemic democratic deficits in formal international governance mechanisms.

The initial target was for a 40 person MAG with 20 members coming from governments and the remaining 20 spots divided between nominated representatives from business and civil society. A 46-person group was eventually selected and raised immediate questions—from civil society in particular—about who was chosen and how these choices were made. The composition of the original MAG, it was argued, created a de facto fourth stakeholder constituency alongside business, government and civil society in privileging the role of the "Internet technical community." This was seen to have occurred at the expense of the more broadly focused or public interest-based civil society actors in areas such as human rights, freedom of expression and privacy. Rough accounting suggests that as many as 9 or more individuals chosen for the MAG had strong and obvious ties to ICANN and/or the Internet Society (ISOC). This whilst, for instance, only 5 names were chosen from the list of 15 nominations that the Civil Society Internet Governance Caucus made through an open and transparent internal process (discussed above). Thus, rather than having 10 members of a 40-member MAG as was the vision at the outset, civil society saw its stake reduced to 7 out of 46. The relative balance of civil society participation was further diminished by the involvement of a group of "special advisors" that the MAG chair asked to assist him and the special invitations that he issued to regional coordinators to attend IGF planning meetings.

As issues that were initially raised by civil society during the consultation process such as free/open source software (F/OSS) and privacy were "refined" out of explicit mention as IGF sub-themes by the MAG, questions were raised about "whether the IGF will actually be the free space for discussion that was

---

14 UN Department of Public Information, *Secretary-General Establishes Advisory Group to Assist Him in Convening Internet Governance Forum.* (May 17, 2006). http://www.un.org/News/Press/docs//2006/sga1006.doc.htm

promised" or, instead, if "the MAG will start to censor themes that are disliked by some of its members."[15] MAG meetings mostly take place behind closed doors and invoke the Chatham House Rule.[16] Thus, neither the decisions themselves nor the agenda-setting role of individual MAG members could effectively be challenged. While certain members of the MAG and some quarters of civil society guarded cautious optimism over their ability to use the workshops to reintroduce these and other themes in a way that would accomplish some bottom-up influence over the IGF agenda, others protested strongly that the power that the MAG seemed to be able to wield over the IGF agenda was incompatible with the WSIS mandate for the IGF. In a message posted on the Civil Society Internet Governance Caucus listserv on May 29, 2006, Carlos Afonso voiced his view that:

> The MAG, an ad hoc creation of the UN Secretary General which is not part of the WSIS decisions, was supposed to help the secretariat regarding IGF procedures, selection process to ensure pluralism and transparency, methodology—not predetermine content or agenda! This should be the task of the first IGF meeting itself—and the IGF should have final say on procedures as well.[17]

In directing their criticisms towards the choices made in forming the MAG, civil society activists lamented the influence that the traditional ills of the UN system—patronage and lobbying, to name two—seemed to have been able to exert on the process of structuring a governance experiment that was, in part at least, premised on addressing the democratic deficit inherent to international policy making.

Immediately following the establishment of the MAG and over the course of the IGF 2006 planning process, concern was expressed about what was signified by the sudden engagement of organizations like ICANN and ISOC that had been defensive towards the WSIS process and vehemently opposed to the idea of the IGF. For instance, in response to the announcement of the MAG's composition, ICANN and Internet governance specialist Milton Mueller protested that:

---

15 Vittorio Bertola, *MAG Dictates Rules and Agenda for IGF?* [WSIS CS-Plenary] Re: [governance]. (May 30, 2006).

16 According to The Royal Institute of International Affairs: "When a meeting, or part thereof, is held under the Chatham House Rule, participants are free to use the information received, but neither the identity nor the affiliation of the speaker(s), nor that of any other participant, may be revealed." http://www.chathamhouse.org.uk/about/chathamhouserule/

17 Carlos Afonso, *MAG Dictates Rules and Agenda for IGF?* [WSIS CS-Plenary]. (May 29, 2006).

the over-representation of direct ICANN agents, via Board members and staff, is troublesome....It's clear that if the results of WSIS did not signal overall acceptance of ICANN's legitimacy and current structure by the intergovernmental system, the initial results of the Forum's MAG selection do.[18]

IGF Institutional Structure and Experience Four Years In

Nitin Desai, the former chair of the WGIG and the UN Secretary-General's Special Adviser for the WSIS, was appointed chair of the IGF. The MAG was created to assume much of the substance and programming responsibility, and a skeleton secretariat was established under the direction of Swiss diplomat Markus Kummer. This intentionally "lightweight" organizational structure was responsible for planning and implementing IGF activities in coordination with local organizing committees established by each year's host country. The IGF and its secretariat are funded through extra-budgetary contributions to a trust fund. The donors are governments, international organizations and stakeholders. These funds are extra-budgetary in the sense that any money the IGF gets must be a donation that goes beyond the usual UN contributions of each stakeholder; there is no way for stakeholders to earmark portions of their existing dues to fund the IGF. The impact is that there is no guaranteed stable source of funding.[19]

The format of the IGF discussions itself has been a source of experimentation over the course of the initial four annual meetings. A two-tracked approach featuring plenary sessions alongside workshops has been a constant.

The IGF is not a traditional decision-making organization, and there are no votes taken or resolutions passed. Workshops are organized by stakeholders based on topics of their own selection. The format of the workshops thus tends to vary, but typically includes a panel of presentations followed by questions, answers and discussion with audience members. Workshops must include participation from each stakeholder group: civil society, government and the private sector.

Plenary or main sessions are typically organized by the MAG around topics and themes that are identified over the course of the consultation process (see table 4). IGF agendas also often include more open-ended plenary sessions devoted to discussion of emerging issues, reporting back on workshops and other parallel activities, and considering the road ahead. During the planning process for the first IGF, it was agreed that "the format of meetings will be

---

18 Milton Mueller. *The Forum MAG: Who Are These People?* http://www.icannwatch.org/article.pl?sid=06/05/18/226205
19 For details on the IGF's benefactors, see The Internet Governance Forum, Funding. http://www.intgovforum.org/cms/index.php/funding

flexible and include moderated panels and discussions both from the floor and from remote participants" with a goal that "each session should be as interactive as possible and devote a large portion of its time to interaction with the meeting attendees."[20] In order to facilitate this sort of dynamic, it was decided that minimal time would be devoted to prepared speeches and that there would be time set aside each day in the main plenary room explicitly for both reviewing the developments from previous days and for open-microphone sessions.

IGF	Overarching theme	Main themes or sub-themes
2006 Athens	Internet Governance for Development	• openness—freedom of expression, free flow of information, ideas and knowledge • security—creating trust and confidence through collaboration • diversity—promoting multilingualism and local content • access—Internet connectivity: policy and cost
2007 Rio	None designated	• critical Internet resources • access • diversity • openness • security
2008 Hyderabad	Internet for All	• reaching the "Next Billion" • promoting cyber-security and trust • managing critical Internet resources
2009 Sharm El Sheikh	Internet Governance: Creating Opportunities for All	• managing critical Internet resources • security, openness and privacy • access and diversity • Internet governance in the light of WSIS Principles

Table 4: IGF Meeting Themes: 2006–2009

At the first IGF meeting in Athens in 2006, these panels each included 14 or so high level speakers from across the stakeholder constituencies. In the attempt to maximize interactivity, each panel consisted of short answers from panelists to questions posed by moderators, typically based on interventions made from audience members. The moderators were professional journalists. In the spirit of the IGF's peer-to-peer multi-stakeholder principles, open-microphone time was allotted on a first-come, first-served basis with govern-

---

20 IGF Secretariat, *Consultations on the Convening of the IGF* (May 19, 2006). http://www.intgovforum.org/meeting.htm

ment representatives having no special privileges regarding speaking time or speaking order. This format received somewhat mixed reviews from participants, with some referring to it as "governance by talk show."

In an effort to make sessions more interactive, to allow for discussion themes and points of discussion to emerge organically from audience interventions, to limit the agenda-setting role played by moderators and to give high level speakers on plenary panels more of an opportunity to meaningfully express themselves, the plenary session format has been refined over the course of the IGF. Moderators are still used to direct traffic, to synthesize discussion and keep it moving and to keep time, but they are increasingly being recruited from amongst IGF participants rather than professional media personalities. Smaller plenary panels have—since the first IGF meeting—been preferred in order to assure that presenters can discuss their views in greater depth. This plenary format assures that—in terms of time allotment and agenda setting—sessions are structured to give significant weight to the interventions of audience members.

Dynamic Coalitions

While the IGF does not pass resolutions, seek to establish consensus or negotiate texts, it provides a platform where stakeholders can coalesce for action around issues of common concern. One of the early MAG strategizing sessions for the first IGF meeting introduced the notion that a series of so-called "dynamic coalitions" each consisting of "a group of institutions or people who agree to pursue an initiative started at the inaugural IGF meeting" could emerge as, themselves, an output of the IGF.[21]

Dynamic coalitions (DCs) have emerged since the first IGF as a way of using the multi-stakeholder platform of the IGF to facilitate coordination, capacity building and awareness raising of Internet governance issues elsewhere. While some of these informal working groups have emerged organically on the basis of discussions held in workshops and plenary sessions at various IGF meetings, others were ready-made for launch at the first IGF in 2006. Some of the DCs formed early on have since collapsed, and new ones have emerged over the course of the IGF. As of the period immediately following the 2009 IGF, the website of the IGF acknowledged the existence of fourteen dynamic coalitions.[22]

---

21 Ibid.
22 Including: Dynamic Coalition on Internet and Climate Change; Dynamic Coalition on Accessibility and Disability; Dynamic Coalition on Child Online Safety; Framework of Principles for the Internet; Gender and Internet Governance; Online Collaboration Dy-

The potential of the IGF to affect Internet governance could be argued to hinge on its ability to provide an institutional platform from which individuals and groups can positively contribute to improving Internet governance. The DC model is thus a compelling governance innovation.

Substantive High Points and Tensions

Amidst fears that political controversy could derail the process before it started, early in the planning process for the initial IGF it was determined that "the" Internet governance issue of the WSIS—political oversight of the ability to make changes to the naming and addressing functions related to the ICANN—would not be on the agenda. A subsequent discussion of ICANN and other DNS-related issues in the critical Internet resources plenary session in Rio was lauded by participants as a necessary step in the evolution of the IGF toward realization of the Tunis Agenda. Given Brazil's role in pushing the issue of internationalized governmental oversight of critical Internet resources during the WSIS, it was perhaps both appropriate and significant that the Rio IGF would in effect re-launch the global debate. Illustrating the evolving capacity of the IGF for confronting politically sensitive issues around the role of governments in the global Internet governance regime, one of the coordinators of the IGC was able to push a discussion of enhanced cooperation onto the 2008 IGF agenda, over objections—from elsewhere within civil society in particular—that the subject was too controversial and risked provoking government disengagement.

The session on critical Internet resources at the 2009 IGF debated the further internationalization of ICANN, the meaning and progress of the enhanced cooperation process agreed to in Tunis was discussed, and what would seem (to certain important governments at least) to be a highly controversial proposal that the IGF should debate the creation of an international body that would take over the Inernet Assigned Numbers Authority (IANA) function from the US government received "strong support."[23] Yet, fear of controversy continues to wield significant pressure within debates about the IGF agenda. Proposals to position human rights and Internet as a main theme of the 2009

---

namic Coalition; Freedom of Expression and Freedom of the Media on the Internet; A2K@IGF Dynamic Coalition; Coalition dynamique pour la diversité linguistique; Dynamic Coalition on the Internet Bill of Rights; Dynamic Coalition on Access and Connectivity for Remote, Rural and Dispersed Communities; Dynamic Coalition on Open Standards; Dynamic Coalition on Privacy; The Stop Spam Alliance.

23 Association for Progressive Communications, *APC's Assessment of the Fourth Internet Governance Forum, Sharm El-Sheikh, 15–18 November 2009.* (November 26, 2009). www.apc.org/en/system/files/APCIGF4Assessment_EN.pdf

IGF were rejected by the MAG as too controversial despite being strongly supported by segments of civil society. Even before that point, however, there were internal debates about this proposal within CS, as certain prominent CS participants cautioned that CS should steer clear of proposing discussion topics that might lead to the disengagement of certain governments.

The issue of online child protection is one notable area where the IGF process has facilitated real momentum toward concrete results and where the true potential of contribution of the IGF format may be emerging. A dynamic coalition on child online safety was created during the 2007 IGF in Rio with the aim of creating "a permanent, open platform for discussion on fundamental and practical issues related to child online safety within the agenda of the Internet Governance Forum, ensuring dialogue among representatives from children's organizations, government, industry, academia and other civil society groups." The contingent of online child protection activists at the 2008 meeting was large and vocal.

As a result, online child protection emerged as something of a cross-cutting issue. Workshops were organized with titles such as "Child safety online: measures to protect children from exploitation—the challenge of keeping pace with technological developments;" "Dignity, security and privacy of children on the Internet—applying international law to protect their best interests;" "Strategies to prevent and fight child pornography in developing countries;" "The Internet goes mobile—child protection in the always connected age;" and "An Interpol for the Internet?" British online child advocate John Carr was given a high profile speaking slot in one plenary. There were numerous interventions made from the audience linking many general topics under discussion back to child protection issues. However, some of these contributions seemed based on more general and rhetorical questions about what was being done to protect children online and exhibited an alarming and largely unproductive degree of instrumental faith in the ability of technical management to weed out undesirable human behavior. Furthermore, not all participants in the IGF were even in agreement that an Internet Governance Forum is the appropriate policy venue for consideration of such issues. In his summary of the Hyderabad meeting, the chair of the IGF felt obliged to observe that: "On child pornography, some people questioned the predominance this topic was taking at this IGF. A number of points were made that this perhaps was not the appropriate space to take up this discussion any further."[24]

---

24 See Chairman of the IGF, *Third Meeting of the Internet Governance Forum (IGF) Hyderabad, India, 3–6 December 2008: Chairman's Summary.* (December 2008). http://www.intgovforum.org/cms/index.php/component/content/article/295-event-in-mumbai

Within civil society, privacy activists in particular were critical of the human rights implications of the invasive surveillance powers that calls for child protection might provide to security and police forces and expressed frustration over what was perceived by some as a coordinated effort at hijacking the IGF agenda. However, other participants seemed to see the emergence, focus and evolution of online child protection within global Internet governance over the course of the 2007 and 2008 IGFs in much more positive terms. Brazilian government representative Everton Lucero suggested that:

> one of the issues that has been debated at length today was the question of child protection against sexual abuse and pornography. And it seems that discussion has matured enough in this area so that now we perhaps could think of creating a common environment where all relevant stakeholders could build trust and work together.[25]

Also noteworthy were the discussions of the Internet governance implications of social media platforms and the consideration given to the implications of increasing mobile phone penetration in the developing world on notions of access. Each is an example of how the policy field of Internet governance continues to broaden, through the multi-stakeholder discussions at the IGF beyond the narrow focus on Internet naming and numbering.

The IGF and Multi-stakeholder Global Governance

Overall, however, the IGF is primarily intended to be a forum for multi-stakeholder dialogue, an institution endowed with the potential for influence and "soft power" rather than an agent in its own right in the international arena. It is also an innovative arrangement set up in a highly contentious area of policy with extremely limited resources that depends on the goodwill and cooperation of its participants. In this respect, even its "soft power" potential was severely constrained from the start.

Through ongoing debate over the setting of agendas, the organization of meetings and the composition and role of its organizing committee, the IGF has continued to refine the process of multi-stakeholder global governance. The IGF multi-stakeholder format has been lauded by many of its loyal participants. It has also been repurposed within various other institutions involved in global communication governance such as the World Intellectual

---

25 Quoted from the transcript of the IGF 2008 "Open Dialogue" session on "Promoting Cyber-Security and Trust." http://www.intgovforum.org/cms/hyderabad_prog/Open%20Dialogue.html

Property Organization (WIPO); the Organization for Economic Cooperation and Development (OECD); and WSIS follow-up activities (see Chapter 8).

The IGF model has been repurposed and adapted for the sub-global level as well. Various regional, national and local multi-stakeholder meetings discussing pressing Internet policy issues have been convened around the world. The model is spreading with more new events being planned each year. The effectiveness of these local IGFs has been lauded by participants and promoted within the UN IGF. In its report on the 2009 IGF, the Association for Progressive Communications (APC) went as far as to argue that the success of these more narrowly focused events suggests that the UN should consider planning thematic IGFs—devoted to more vertical discussion of individual issues—in parallel to the existing larger, more comprehensive, global IGF.[26]

As a venue for meaningfully continuing the WSIS debate over Internet governance, however, the IGF model of non-binding multi-stakeholder governance has proven to be a source of frustration for governments bent on changing the structures of IG. "It's not by talking about principles merely that we can solve this problem," the Chinese delegation intervened at a consultation on the review of the IGF's mandate, continuing that

> China does not agree with extending the mission of the IGF beyond the five years. We feel that after the five years are up, we would need to look at the results that have been achieved. And we need, then, to launch into an intergovernmental discussion.[27]

Despite what the IGF has accomplished as an experiment in global governance and as a forum for sharing best practices and discussing emerging issues, stakeholder frustration is understandable. For starters, many developing countries lack the resources and capacity to meaningfully participate in the IGF. Furthermore, the dynamic coalitions have proven to be difficult to sustain. Face-to-face meetings outside the annual IGF event are cost prohibitive. The participation of government delegates is complex and problematic given that civil servants face both time constraints and ethical challenges related to their ability to contribute in such a format. But in spite of limited impact of many of the DCs, over the course of its first four years the IGF process has incubated multi-stakeholder responses to a handful of important Internet governance policy issues.

---

26   Association for Progressive Communications, *APC's Assessment of the Fourth Internet Governance Forum, Sharm El-Sheikh, 15–18 November 2009*. (November 26, 2009). www.apc.org/en/system/files/APCIGF4Assessment_EN.pdf

27   The Internet Governance Forum website, *Transcript of the 13 May Open Consultations Full Transcript*. (May 13, 2009). http://www.intgovforum.org/cms/index.php/component/content/article/71-transcripts-/410-transcript-of-the-13-may-open-consultations-

Reflecting the tensions observed in the DCs, the IGF in general suffers from a lack of meaningful engagement on the part of many government delegations. This in turn has led to a situation in which other stakeholders, in particular civil society, are engaging in what amounts to self-censorship in the effort to frame discussions in a way that will attract government engagement and prevent government disengagement.

More important perhaps, the IGF fails to demonstrate the levels of coordination and sufficient political capital required to influence other organizations. Many of the dynamic coalitions, for example, seem largely confined to meeting at the annual IGF event. In regard to its mandate to help build capacity and encourage developing country participation in global Internet governance, raising funding to provide fellowships for developing country participants has been a challenge and remote/online participation has been unreliable, inconsistent and largely underused.

Fundamentally, with the exception of certain areas (protection against online child pornography and multilingualism stand out as issues where the IGF has been a valuable venue for concrete, coordinated multi-stakeholder activity), the IGF remains preoccupied with questions of process rather than substance. The dominant topics include debate about agenda setting, the role of various stakeholders and the status and constitution of its multi-stakeholder advisory group. The upcoming five-year review of the IGF mandate called for in Tunis and the review process that is already underway threaten only to accentuate this tendency. As a result, the IGF has been very slow to get off the ground in building any kind of real momentum.

Given the middling impact of the other elements of WSIS follow-up, dissolution of the IGF would lead to serious questions about long term impact, in concrete institutional terms, of the 5 years and millions of dollars that were invested in the WSIS process. As other international organizations such as the OECD and ITU expand their involvement in global Internet governance, and as reform-minded governments critique and disengage from its non-binding modalities, the IGF faces a serious threat from organizational competition just as its mandate is being reviewed, despite its status as a still-evolving provocative and innovative model of global Internet governance.

Venue shopping and the (re)emergence of the ITU threat to multi-stakeholderism

The push to position the ITU as an intergovernmental Internet governance organization that dominated the first phase of the WSIS is arguably re-emerging. Mandated by a resolution passed by the 2006 ITU Plenipotentiary

Conference instructing him to take a "significant role" in the management of Internet and DNS resources, the coordination of public policy issues pertaining to the Internet and the process of enhanced cooperation,[28] ITU Secretary General Hamadoun Touré made very public scathing critiques of the ineffectiveness of the IGF and its non-binding modalities to a November 2008 ICANN meeting in Cairo. "I personally believe," Touré offered, "that the IGF is just going around and around, avoiding the topics, and becomes sometimes a waste of time." [29] During the opening session of the 2008 Hyderabad IGF, Touré addressed his Cairo comments, insisting

> I make no apology for stating bluntly that I believe the IGF was not on track to meet the expectations of many countries that participated in the Tunis phase of WSIS [...] who were hoping for frank and fruitful discussions and concrete solutions on globally applicable principles for the management of critical Internet resources.[30]

It is likely little coincidence that the ITU is planning to host a World Conference on International Telecommunications (WCIT) for what would be year 6 of the IGF's existing five-year mandate. If the IGF mandate is not renewed, the WCIT will be well positioned to fill the space vacated by the IGF. The institutional competition that defined the WSIS debates over ICANN vs. ITU is likely to continue with the ITU bearing the standard for the ambitions of those governments who favour intergovernmental Internet governance. The agenda being pursued by various governments in the ITU, in other words, is openly hostile to the multi-stakeholder Internet governance model. It seems highly unlikely that the same governments would approve of significant reforms to the ITU that aimed to import the multi-stakeholder model into intergovernmental processes (see Chapter 8). Despite the in-roads that discourses about multi-stakeholder global governance have made in the area of Internet governance, the continuation of CS participation in such policy debates remains threatened on various fronts.

---

28 ITU, *Resolution 102 (Rev. Antalya, 2006)ITU's Role with Regard to International Public Policy Issues Pertaining to the Internet and the Management of Internet Resources, including domain names and addresses*. (2006). http://www.itu.int/osg/csd/wtpf/wtpf2009/documents/ITUresolution 102_ publicpolicy_IPbasednetworks_PP06.html

29 ICANN, ICANN *Meeting–Cairo Thursday, 6 November 2008: Hamadoun Touré Speech*. (November 6, 2008). https://cai.icann.org/files/meetings/cairo2008/toure-speech-06nov08.txt

30 IGF, *Internet Governance Forum: Hyderabad, India Opening Session 3 December 2008*. (December 3, 2008). www.intgovforum.org/cms/hydera/Opening%20Session.pdf

The Multi-Stakeholder Model:
Reflections from the Internet Governance Experience

Civil society actors engaged in WSIS discussions on a variety of issues and came to them from a variety of perspectives. During the first phase of the WSIS, Internet governance was one among a constellation of issues being considered through the emerging multi-stakeholder governance model. The debate changed during the second phase, becoming more focused on a specific issue: Internet governance. Civil society participation changed too. Groups and individuals who had been important CS voices and held influential offices in the CS structures that had been created during Phase I took a step back or dropped out all together over the course of Phase II. Certain actors' issues fell down or off the WSIS agenda at the conclusion of Phase I, resources ran out, other projects and commitments intervened, process fatigue and declining interest set in.

One result was that the multi-stakeholder global governance experiment was pushed furthest in the area of Internet governance. It is thus worth reconsidering here what has been accomplished, sacrificed and, above all, learned about multi-stakeholderism from the global Internet governance experience.

Multi-stakeholder model reviews: The shift from tripartite to multi-stakeholder to...what?

Multi-stakeholder discourses during the WSIS centered on a tripartite notion of global governance, wherein the private sector, governments and civil society contributed—in their own roles—as separate sectors. Over the course of the initial IGF meetings there was considerable enthusiasm for the idea that this had or could evolve into a multi-partite model—wherein all delegates were co-equal participants in the multi-stakeholder process, regardless of their stakeholder constituency.

The experience of the initial IGFs, however, underlined some of the political limitations of this approach. For one, there was constant tension around the participation of government delegations and individual government delegates in such arrangements, as state interests and personal careers could potentially be jeopardized by any missteps civil servants might make. Traditional intergovernmental policymaking, plodding, rule-intensive though it may be, assures that government delegations speak in a singular voice that is coordinated with the decision-makers back home. The sort of collective brainstorming that occurs at the IGF seems to require individual civil servants who want to engage meaningfully to commit to sharing positions and information with no guarantee of reciprocity and, in many cases, to improvising on the spot responses to questions and new issues that may not even have been anticipated, much less vetted.

As a result, individual civil servants' engagement in the IGF has proven to be a challenge. By the third IGF, the interventions of a series of OECD countries and various private sector representatives seemed to be evidencing a strong endorsement for an emerging revisionist take on multi-stakeholderism.[31] This view ascribes real value to multi-stakeholder engagement on a very practical, case-by-case basis which puts the emphasis on best practice and recognizes the need for "variable geometry"—i.e. the idea that the roles and responsibilities of the different stakeholder groups will vary according to the issue, the kinds of practical responses it requires in terms of policy, regulation, innovation, awareness raising, etc., and economic, social and cultural contexts. In Hyderabad, this view was discussed by the US government's Dick Beaird, perhaps most concretely articulated by British Member of Parliament Alun Michael in his various interventions throughout the event,[32] and most logically articulated by French government delegate Bertrand de La Chapelle in the open dialogue on critical Internet resources.[33]

As Dick Beaird explicitly said, this is a very different view of multi-stakeholderism from the one that arose out of WSIS and is embedded in the summit's official documents. The revisionist multi-stakeholder model is more top-down in its orientation and sees the challenge of multi-stakeholderism essentially as the challenge of defining the roles and responsibilities of government, the private sector and civil society in Internet governance organizations and decision-making processes, primarily at the international level. Recognizing that there will be variable geometries in these forums as well as in practical reality, the main general problem that needs to be addressed with respect to this point of view for civil society is the need to obtain a seat at the table and have a voice in decision-making processes. This viewpoint was well articulated at the 2008 IGF by Parminder Jeet Singh and Milton Mueller in the panel session on global, regional and national arrangements for Internet govern-

---

31 Conversation with with Ottawa-based ICT policy consultant Don MacLean influenced much of this section and many of the observations and terminologies are properly credited to him.
32 See the Transcripts of the IGF 2008 sessions on the IGF website at http://www.intgovforum.org/cms/index.php/2008-igf-hyderabad/hyderabadprogramme
33 Internet Governance Forum, *Hyderabad, India. Managing Critical Internet Resources Open Dialogue.* (December 5, 2008). http://www.intgovforum.org/cms/hyderabad_prog/OD_CIR.html

ance,[34] and by Bill Drake in his proposal for a workshop and dynamic coalition on a development agenda for Internet governance.[35]

Reflections on CS in Internet Governance

The Ideal Role

Most governments engage in some form of consultation with stakeholders on the positions that they adopt in global governance decision-making processes. These consultations would ideally involve citizens in the adoption of various positions. In reality governments typically seek out the opinions of industry or expert specialists to help fill in knowledge gaps and engage with key domestic firms and industry lobbies in the effort to determine a course of action. This may have benefits for local economic development and prosperity or it may serve the interests of the private sector alone.

Thus, the ideal role for civil society participation in global governance includes building the capacity to counteract the lobbying efforts of the private sector through the creation of relationships and alliances with policy makers and integration of CS voices into governmental consultation and policy development processes. CS participation also contains a presumptive check on governance by invisible, backroom deals that are not accountable to public scrutiny. It can also serve to counter the more authoritarian tendencies and practices that all governments exhibit at times, some more frequently and more ferociously than others. Through their own grassroots networks, communication platforms and information dissemination practices, CS actors can solicit and impart the input of a wider public opinion and the voices of stakeholders who are marginalized by the usual channels of discourse in intergovernmental politics. Pragmatically, CS participation also allows for the integration of valuable specialist knowledge into the policy development cycle and creates a standing reserve of in-house specialists whose expertise can be easily referenced by decision-makers during negotiations.

---

34 Reference to the comments of Beaird, Singh and Mueller discussed here can be found in the transcript on global, regional and national arrangements for Internet governance. Internet Governance Forum, *Hyderabad, India. Arrangements for Internet Governance, Global and National/Regional. 5 December 2008.* (December 5, 2008). http://www.intgovforum.org/cms/hyderabad_prog/AfIGGN.html

35 See Graduate Institute for International and Development Studies, *A Development Agenda for Internet Governance: From Principle to Practice.* (November 14, 2007). http://www.intgovforum.org/cms/workshops_08/showmelist.php?mem=15

Each of these roles has been played at times by various CS actors over the course of the global Internet governance debate. But, in truth, CS participation in Internet governance has not always conformed to the ideal.

Critiques

The debate on global Internet governance is itself preoccupied with process over substance. From roles and responsibilities of governments and other stakeholders, to the question of oversight and the discussion of the IGF, the WSIS debate was largely—though not exclusively—a debate about what could be described as the institutional process in which Internet governance occurs. In the IGF and other venues, debates about institutional structures, processes and working modalities have dominated agendas.

Thus, it is not surprising that this debate tended to attract and produce prominent CS voices that were also focused on process. Additionally, CS positions on Internet governance were diverse and CS was typically unable to speak with a consensus view on Internet governance issues.[36] The notable exception was the role of CS itself within the institutional structures of global Internet governance. Over the course of the second phase of the WSIS, the continued involvement of CS and the general push toward multi-stakeholder global governance of the Internet became something of a rallying cry for CS positions. The idea that CS must play a fundamental role in global Internet governance has continued as a dominant theme of discussions of and at the IGF and in regard to new initiatives that have emerged.

Yet, at a certain point, the question has to be asked: to what ends? If civil society participation is going to focus on the processes through which civil society participation can be assured and improved upon, at what point does the cart of substantive results follow the horse of multi-stakeholder process?

Certainly, many individual CS participants have impacted various debates on global Internet governance with important substantive political projects. However, five years into the IGF experience, the frequency and voracity of calls that global Internet governance be multi-stakeholder and open to CS leads naturally to questions about why and about what, of substance, has been accomplished thus far with the access that CS has been granted to global governance decision making over the course of the WSIS and IGF.

---

36 Adam Peake, *Internet Governance and the World Summit on the Information Society (WSIS). An Issue Paper of the Assocation for Progressive Communications.* (June 2004). http://www.apc.org/en/pubs/issue/governance/all/Internet-governance-and-world-Summit-information-s

In part, the ongoing push for process impact rather than substantive impact reflects on the problematic role of CS delegates who represent only themselves as individuals in contrast to those who participate as part of the delegation of an NGO or other form of network or organization. The participation of individual CS delegates brings with it the potential for individual gain. The access to the corridors of power that is granted through multi-stakeholder governance arrangements creates opportunities for networking, for increasing personal status and reputation. Professional consultants gain expertise and contacts that lead to consulting contracts, and academics get access to privileged research terrains (that can, evidently, lead to publishing contracts). The highly context-specific sort of specialist knowledge acquired by dedicated individual CS participants is extremely valuable to larger organizations who are also engaged in the process, and many individual WSIS CS delegates have since taken full time positions or consulting jobs with some of the governments, private sector entities, international organizations and NGOs that remain at the centre of the global Internet governance debate.[37] In this sense, it is problematic to try to disaggregate the push for greater civil society involvement in global communication governance from the potential personal gains that such a move would provide to many of the individual CS delegates at the centre of the push.

The professionalization of CS participation in global governance processes is thus in some respects the elephant in the room of this discussion. Important CS organizations as well often seem to get caught up in professionalized lobbying activities, but the drive for more opportunities for CS participation in global Internet governance is far less problematic when the push emanates from representative, broad-based not-for-profit groups with secure and sustainable resources. A shift from a network of individuals to a coalition of groups would mean that CS delegates had less at stake individually from the multi-stakeholder processes in which they participate.

A knock-on effect of this tendency is the problem of CS enrollment in the multi-stakeholder processes that have emerged since WSIS. In particular, CS participation in the IGF shows evidence of a tension between cooptation and integration. The IGF is often credited as being a creation of CS members of the WGIG, CS pushed hard for its creation during the WSIS, and a number of CS actors have served on the MAG and have otherwise invested a good deal

---

37 Of course, this cross pollination is not unique to CS, as many of the key government delegates have also since taken up new positions, in some cases with large key non-governmental stakeholders who value their contacts and familiarity with government policy development and intergovernmental politics. Moreover, needless to say, this phenomenon is not exclusive to the Internet governance sector.

of time and resources in making the IGF work. In many respects, the fortunes of the IGF will reflect strongly on the future of CS inclusion in global Internet governance.

Yet the need for CS to integrate into the practices and political realities of global Internet governance frequently leads to calls that CS muffle its voice. At what point does integration become cooptation? The line is difficult to situate precisely, but there are numerous examples that seem to suggest it is being approached, if not crossed. In particular this tension manifests itself as pressure that certain CS actors put on other CS actors to step back from using the IGF as a venue for discussing and pushing an activist agenda around issues that might be controversial to certain governments on issues such as human rights and critical Internet resources. If governments disengage, the logic goes, the IGF suffers, and if the IGF fails, all CS suffers. The parameters for action become limited and the end goal gradually shifts toward keeping the CS-friendly process alive rather than achieving the substantive results desired.

The global Internet governance debate also reveals existing tensions between CS and other stakeholders. This was notably the case between the private sector and CS, with some CS delegates operating under normative critiques of the private sector and refusing to collaborate under any circumstances. But there were also, to a certain degree, culture clashes that occurred between government officials and CS delegates. Over the course of the WSIS there were tensions and conflicts within civil society itself as constituencies battled each other to get their issues on the political agenda, there were conflicts between various caucuses and working groups, and within certain caucuses, the IGC in particular. On certain issues, CS simply had to deal with competing well-intentioned perspectives on the appropriate policy response. This was the case, for example, with the tensions between privacy activists and child protection advocates that emerged at the 2008 IGF. At other moments, however, the experience of CS internal relations in regard to the global Internet governance debate raised fundamental questions about how CS participation, given its openness to anyone, and its need to provide some sense of a cosmopolitan basis to a global constituency that spans many cultures (not to mention first languages) may simply suggest that such structures are difficult to administer and coordinate. Also of concern should be the barriers to entry for new participants created by process-intensive language and demanding organizational culture of certain spheres of CS activity (the IGC listserv for example).

What CS inclusion accomplished in global Internet governance

On balance, has the participation of CS in the global debate over Internet governance contributed to improvement in the governance of the Internet?

Between April and June 2009, the National Telecommunications and Information Administration (NTIA) of the US government's Department of Commerce (DOC) solicited public comments on a range of questions about whether or not ICANN has evolved to the point where the US government could end its oversight of ICANN by allowing the controversial relationship between the two that had been codified by various Joint Project Agreements (JPA) and Memoranda of Understanding (MoU) to expire.[38]

These consultations reflect how the Internet governance debate has evolved over the course of WSIS Phase II and since. A series of non-US based individuals, and civil society organizations including German academic Wolfgang Kleinwächter, South African-based NGO the Association for Progressive Communications (APC), Indian NGO IT for Change and the global network of activists in the WSIS Civil Society Internet Governance Caucus all submitted comments to an American government policy consultation. Comments were also received from foreign government departments including Industry Canada and the Swiss Federal Office of Communications.[39] This latest round of debates within the US on the ICANN MoU underlines the extent to which the debate on IG has globalized and to which the WSIS has incubated an international, multi-stakeholder policy community around Internet governance. Its outcomes also reinforce how the participation of CS in these debates is contributing to a greater focus—at the rhetorical level at least—on transparency and accountability and, in particular, on the public's interest within the global debate on Internet governance.

On September 30, 2009, the US Department of Commerce and ICANN announced the policy framework that would succeed the series of MoUs that had defined oversight of global Internet governance since the 1990s. Oversight of ICANN and the nature of the agreement that would replace the MoU had, as we have discussed, been the focus of much of the debate during the second phase of the WSIS. The new agreement, called the "Affirmation of Commitments by the United States Department of Commerce and the Internet Corporation for Assigned Names and Numbers"[40] reaffirms the role of

---

38 NTIA, *Federal Register/Vol. 74, No. 78/Friday, April 24, 2009/Notices*, NTIA. (April 24, 2009). www.ntia.doc.gov/frnotices/2009/FR_ICANNVol.74_No78_Apr242009.pdf
39 NTIA, *Public Comments: DNS Transition*. http://www.ntia.doc.gov/comments/2009/dnstransition/index.html
40 See ICANN, *ICANN CEO Talks About the New Affirmation of Commitments*. (September 30, 2009). http://www.icann.org/en/announcements/announcement-30sep09-en.htm

governments in developing and reviewing ICANN decisions, making token gestures toward greater independence from the US government and internationalization without addressing the role of US oversight of the IANA function. In other words, in terms of what it does, the Affirmation of Commitments does not particularly reflect the extent to which civil society participated in the eight-year debate over global Internet governance that the WSIS incubated and the Affirmation, in part, responded to. Yet the term "public interest" is mentioned five separate times in this 2000-word document. There are also numerous terms containing the word "public" that seem to stand in for, or go into greater detail on some element of the public's interest in global governance of Internet domain names and addresses. For example: "effects on the public;" "public policy aspects;" "public input;" "public comment;" "public access;" "impact on the public;" and the idea that "decisions should be...accepted by the public." Many of these appear multiple times in the document; there are 26 such "public interest" related notions mentioned in total.

On the face of it, it isn't extraordinary that discourse around the public interest in communication is prominently featured in an important piece of communication policy.[41] Viewed in a historical context, though, what is remarkable is that the public interest features so prominently in an important piece of *Internet* policy. The very notion of the public interest in communication as a primary decision-making guidepost for policymakers presupposes certain characteristics of a communication system, namely that regulatory intervention or oversight of communication in some form is both possible and desirable.

Until recently, the tendency has been to treat the Internet as a form of communication that failed each of these tests. Policy frameworks developed in the late 1990s around the world supported this view, as did courts, scholars from a variety of disciplines and technologists. Internet regulation, according to this conventional wisdom was

- impossible: because Internet mediated communication is borderless—the receiver of information is not necessarily physically located within the same jurisdiction as its sender—it has been perceived to create intractable conflicts between existing national legal frameworks, in particular in regard to the definition of what constitutes acceptable limits on freedom of

---

41 The public interest is usefully defined as "the primary decision-making guidepost for policymakers" and "the primary criterion against which policies are assessed." See Philip M. Napoli. *Foundations of Communications Policy: Principles and Process in the Regulation of Electronic Media*. Cresskill, NJ: Hampton Press Inc, 2001 (at 72).

expression. In a text published in 1996, legal scholars Johnson and Post wrote: "Any efforts to map local regulation and physical boundaries onto cyberspace are likely to prove futile."[42]

- undesirable: because many argued that Internet technologies and communities were self-governing and that public interest values such as freedom of expression, community and democracy were embedded into the technologies themselves. "A world more humane and fair than the world your governments have made before," one infamous proponent wrote in 1996.[43]

Against this backdrop, a series of standards-making bodies, populated by engineers and technologists and orchestrated by a core group of Internet "founding fathers," were meanwhile making decisions about the functioning and development of the Internet. In this sense, the Internet was less unregulated than it was self-regulated through this "private technical management."[44]

ICANN, in many respects, exemplified these trends. Private technical management was typically framed as a utilitarian function, with efficiency as the end goal rather than the public interest. Indeed, in the first MoU between the US DoC and the ICANN from 1998—essentially a previous version of the Affirmation of Commitments—the term "the public interest" does not appear in the text. The purpose of this statement of policy is defined entirely in market based terms:

> the privatization of the technical management of the DNS in a manner that allows for the development of robust competition in the management of Internet names and addresses.

The only public interest-related phrase that appears in this document is the idea of "mechanisms to solicit public input...into a private sector decision making process."[45] To the extent that the post-WSIS document mentions the public interest 26 times, the Affirmation of Commitments arguably reflects

---

42 David R. Johnson and David G. Post, "Law and Borders: The Rise of Law in Cyberspace." *Stanford Law Review* 48, 1996 (at 1367).
43 John Perry Barlow. *A Declaration of the Independence of Cyberspace*. (1996). http://homes.eff.org/~barlow/Declaration-Final.html
44 See Milton Mueller, *Ruling the Root: Internet Governance and the Taming of Cyberspace*. Cambridge, Mass: MIT Press, 2002; Daniel Paré, *Internet Governance in Transition: Who is the Master of this Domain?* Lanham, MD: Rowman & Littlefield, 2003; Jack Goldsmith and Tim Wu, *Who Controls the Internet? Illusions of a Borderless World*. New York: Oxford University Press, 2006.
45 United States Department of Commerce, *Management of Internet Names and Addresses*. http://www.ntia.doc.gov/NTIAhome/domainname/6_5_98DNS.htm

the involvement of civil society in the WSIS and in global governance debate since, if not in terms of what it does, then, in at least in terms of what it says.

The scrutiny of civil society has embedded within global governance structures the principle that decisions taken must be at least nominally justified as being in the public interest. The expertise of civil society actors who have participated in the WSIS debates and since has, for example, contributed to a process of collective learning wherein government delegations have come to a more sophisticated and informed understanding of the functioning of Internet technologies and governance processes and has facilitated a shift in conventional wisdom about the prospects for and desirability of control and governance of communication over the Internet.

The contribution that the participation of civil society in the WSIS and other multi-stakeholder governance structures has made to injecting some semblance of the public interest into the global governance of the Internet cannot be ignored. In the following chapter, we will map out some of the impacts of multi-stakeholderism on the global governance of communication that have expanded beyond the WSIS-IGF nexus.

• CHAPTER EIGHT •

Post-WSIS Civil Society Engagement

In the aftermath of the WSIS, civil society participation in the global governance of communication has expanded and evolved beyond just the IGF. The majority of the structures erected to coordinate civil society participation in the WSIS were taken down with its conclusion. Certain CS organizations created for the WSIS have endured in some form and turned their attention to other venues and activities. By the end of the first phase of the WSIS, the CRIS campaign, for example, was turning its attention to other issues and institutions. A few months after the conclusion of the Geneva phase of WSIS, a posting from a CRIS activist summarized the campaign's next steps as follows:

> CRIS will continue to focus on issues systematically ignored by the WSIS, while emphasizing the need for civil society to engage with the real poles of power in global governance of communications: the 'free trade' regime of WTO [World Trade Organization], FTAA [Free Trade Area of the Americas], and other regional and bilateral trade agreements, as well as venues like WIPO and ICANN, where communication rights face growing threats, and UNESCO where the Cultural Diversity Convention [on the Protection and Promotion of the Diversity of Cultural Expressions] demands our urgent support and reinforcement.[1]

Most fundamentally, the impact of the WSIS on civil society participation in the global governance of communication has been the contributions that have proved portable to other venues: experience and connections. Over the course of the WSIS, CS activists developed networks and personal relationships not only with other CS activists, but with key delegates and representatives from various governments, international organizations and the private sector as well. In addition, the WSIS gave invaluable experience in the processes, politics and machinations of institutional global governance to the individual CS actors and organizations that participated.

This chapter will reflect on the evolution of civil society participation in the global governance of communication by briefly reviewing some of the ways in which the multi-stakeholder global governance model is spreading beyond

---

1   Sasha Costanza-Chock, CRIS and WSIS Phase II. [income]. (June 22, 2004).

the WSIS to an institutional cluster that includes: WSIS follow-up and implementation structures, the World Social Forum, ICANN, WIPO, UNESCO, ITU, the Global Alliance for ICT and Development (GAID) and the OECD.

WSIS Implementation and Follow-up

CS participation in and satisfaction with the various WSIS follow-up and implementation activities have been uneven. Despite the concerns voiced during the WSIS that the lack of a binding mandate for follow-up and implementation activities would allow governments to disengage post-WSIS, government participation has been strong in the stocktaking activities at least (where it is most easily measurable and involves the least commitment). The 2008 ITU report on WSIS stocktaking puts government inputs to the process at 54% of the contributions received compared to 11% from CS.[2]

In the aftermath of the WSIS, an interesting unofficial, parallel effort to monitor and analyze trends and initiatives in the development of the global information society has emerged from a coalition of CS groups led by the Association for Progressive Communications (APC). Critical of limits of the narrow and strictly quantitative measures developed by the ITU, the APC-sponsored Global Information Society Watch presents an annual report that uses a mix of indicators and qualitative analysis to examine a handful of cross-cutting issues, map out recent institutional activities relevant to the information society and record country reports on the situation on the ground.[3] Its explicit focus is ICTs for development (ITCD).

CS was, as we have discussed (see Chapters 2 and 6), unsure about what to expect from the action line facilitation process. According to one CS participant, Willie Currie from APC, the first round of action line facilitation meetings included:

- a report on WSIS outcomes in the respective area of the respective action line;
- briefings by participants on their respective projects;
- presentations by stakeholders on possible priorities for action and modalities for cooperation;

---

2 See the 2008 ITU Report on WSIS Stocktaking. http://www.itu.int/wsis/stocktaking/index.html
3 See the Global Information Society Watch (GISWatch) 2009 Report at: http://www.giswatch.org/

- exchange of views by participants on the objectives of the group.[4]

Seen in the flesh, the evaluation of the abstractly devised, but crucially important element of WSIS follow-up was less than kind. In the assessment of ICTD consultant and scholar David Souter,

> the first round of "action line" meetings held in May 2006 was very poorly attended and produced little in the way of new initiatives...it is difficult to see the action line structure, which has no independent resources, offering much of a framework for future cooperation or any significant legacy for the WSIS.[5]

In subsequent years, WSIS follow-up meetings have been held as a part of a "cluster" or "forum," where CSTD meetings, IGF MAG meetings and open consultations and other WSIS related events are held consecutively with action line facilitation meetings.

In comments delivered to a May 2009 meeting of the CSTD, APC executive director Anriette Esterhuysen said that the experience in participating in the panoply of WSIS follow-up activities suggests that "they can be immensely valuable, but only if they create concrete linkages between the global and the national levels; between people; and between discussion and action." From there, however, Esterhuysen's remarks on the role of CS within WSIS follow-up turn more critical:

> Within the context of international WSIS-follow-up many platforms and events have included multiple stakeholders, but, only the IGF (Internet Governance Forum) has systematically enabled non-governmental actors to be involved in agenda-setting. It is also, thus far, the only forum that has produced self-organized regional and national multi-stakeholder sister-forums.[6]

Overall, however, these cross-cutting comments from one of CS's most prominent CS organizations underline the reality that WSIS follow-up and implementation outside of the IGF has thus far failed to mobilize CS to any significant extent or hold the attention of many CS actors who participated in the WSIS. However, it will be difficult to gauge the impact of these elements

---

4 See Willie Currie, "Post WSIS Spaces." in *Global Information Society Watch 2007*. (2007 at p. 18). http://www.giswatch.org/gisw2007/

5 See David Souter, "The World Summit on the Information Society: The End of an Era or the Start of Something New?" In *Global Information Society Watch 2007*. (2007 at p. 15). http://www.giswatch.org/gisw2007/download

6 Remarks presented on the opening panel of the 12th session of the United Nations Commission on Science and Technology for Development by Anriette Esterhuysen Executive Director of the Association for Progressive Communications—25 May 2009. (audio) http://www.unctad.org/sections/meetings/audio/2009-05-25/am/1012-E.MP3

of WSIS follow-up and the contribution that CS has made to them until formal analysis of their impacts are undertaken for the WSIS review conference slated for 2015.

UNESCO

Negotiations at UNESCO in 2004 and 2005 over the adoption of a Convention on the Protection and Promotion of the Diversity of Cultural Expressions mobilized a network of activists and civil society organizations.[7] The CRIS campaign, in particular, invested heavily in this debate, working to raise awareness of the issues related to the Convention, distributing timely updates on the progress of negotiations and drafting policy statements, and otherwise attempting to influence the negotiations taking place at UNESCO.

The issues and themes covered by the negotiations regarding the UNESCO Convention were broad, but largely centred around the following elements:

- counteracting the trend to reduce cultural products and expressions to mere commodities in international trade agreements;
- affirming the right of states to develop and implement cultural and communication policies;
- providing legal legitimacy to and support for cultural and communication policies;
- protecting states who maintain cultural and communication policies from being penalized under current or future international trade agreements;
- avoiding the subordination of culture to the logic and requirements of free trade and commodification.

CRIS added its voice to the efforts of other CS groups mobilized around the draft Convention such as the International Network for Cultural Diversity (INCD) and Free Press and requested the following from the agreement:

> First, the Convention must not be made subordinate to existing or future trade agreements. To do so would defeat its purpose.
>
> Second, it should be designed to not only protect diversity of national and regional cultural industries, but to protect the cultural diversity and the communication rights of all peoples.

---

7 See notably, International Network for Cultural Diversity, http://www.incd.net/incden.html. See also, Coalition for Cultural Diversity: http://www.cdc-ccd.org

Third, it must balance any references to the protection of intellectual property rights with reference to protection of the cultural commons. Otherwise, references to intellectual property rights should be removed altogether.[8]

Civil society groups stayed mobilized around the Convention well after its quasi-unanimous adoption by the General Conference of UNESCO on October 20, 2005. From that point on CS groups focused on the effort to lobby individual governments to take the next step and ratify the Convention to make it an international binding agreement. These efforts were ultimately successful and the Convention came into force on March 18, 2007.

World Social Forum

The normative commitment to democratizing global communication and to creating greater awareness of the role that communication plays in the more general democratization of society at the core of the CRIS campaign led logically to engagement in the World Social Forum process.[9] CRIS involvement in the WSF aimed at sharing the knowledge that had been acquired at WSIS about strategies for participating in, organizing around and influencing global institutional governance structures; discussing issues relating to media, culture and communication, and broadening the foundations of a grassroots movement on communication rights.

CRIS members participated in the first Information and Communication World Forum (ICWF), held in Porto Alegre, Brazil, on January 25, 2005. The event was an open space for those concerned with issues related to culture, information and communication. Its aims were to facilitate discussion on these themes, link them to the global social justice movement and establish a base within the World Social Forum that would serve as a platform for public education and discussion on the right to communicate. According to one report from the Forum:

---

8   Sasha Costanza-Chock, URGENT: Support CRIS Comments on UNESCO Cultural Diversity Convention. [nettime]. (November 11, 2004).

9   Conceived in opposition to the role that the World Economic Forum is perceived to play in coalesciing the ideological underpinnings of neoliberalism, the World Social Forum is a movement of movements, a gathering place for reflection, consensus building and networking amongst social justice-centred civil society groups from around the world. Its general aim is the promotion of people-centred global governance alternatives to neoliberalism. See the charter of the 2002 WSF meeting at http://www.portoalegre 2002.org/default.html

At WSIS civil society allied with communication researchers as the CRIS campaign (Communication Rights in the Information Society), launched by a Communication Rights platform created by a group of NGOs working in the information and communication field. The CRIS campaign fought for citizens in a high-level world conference, which was geared above all to the business aspect of the Information Society. CRIS' effort will continue next year in Tunis, where the follow-up to Geneva will take place. The ICWF will invite CRIS and others to share their testimony, opening an interesting debate.

It needs to be emphasized that the CRIS alliance has highlighted that the right to information has to give way to a new updated phase of the debate: the right to communication. This right opens space and participation to all citizens, and goes beyond the vertical information system, which has given us a few information producers, directing their products to a vast audience, and creating insoluble barriers to citizens' participation. New Information and Communication Technologies (ICTs), which go beyond Internet, open for the first time the possibility to construct a wide horizontal system, participatory and democratic, a communication system at the service of humanity and not of the markets.[10]

The presence and participation of CRIS at the World Social Forum was thus instrumental in positioning the notion of the right to communicate as a catalytic concept around which to build a movement focusing on issues affecting the media, intellectual property rights, access to knowledge, information and communication and culture, and human rights.

ICANN

Part of ICANN's effort to respond to the pressures, critiques and threats that it faced over the course of the WSIS included the commissioning of an independent review of its Generic Names Supporting Organization (GNSO).[11] One of the issues underlined in the commissioned report produced by the London School of Economics (LSE) Public Policy Group was the need to reform the structures and procedures related to the participation of constituencies. Over the course of 2006-2009, ICANN embarked on a process of "GNSO Improvement." A proposal for a Non-Commercial Users Constituency (NCUC) group, framed as "the home for civil society organizations and individuals in the Internet Corporation for Assigned Names and Numbers

---

10   Inter Press Service News Agency, *First Information and Communication World Forum.* (January 25, 2005). http://ipsnews.net/new_adv/cworldforum.asp

11   Simon Bastow, Patrick Dunleavy, Oliver Pearce, and Jane Tinkler, "A Review of the Generic Names Supporting Organization (GNSO)." LSE Public Policy Group, London School of Economics and Political Science, London, UK. (Unpublished).

(ICANN) Generic Names Supporting Organization (GNSO)" was developed and refined.[12] Despite what were perceived to be efforts on the part of ICANN staff to dilute, sideline or otherwise set a top-down agenda for the role of civil society within the new GNSO structures, the charter and structure that had been developed bottom-up by civil society itself was eventually accepted for integration into the GNSO by the ICANN board. As a result, according to one of the key organizers of the NCUC efforts, civil society is "now a much stronger and more active force in global governance of Internet name and number resources."[13]

The overlap between WSIS CS and the NCUC experience is multi-layered and significant. In the first place, some of the central figures in the leadership and membership of the NCUC effort were also prominently involved in WSIS CS structures.[14] The engagement of many of these individuals in ICANN pre-dates their participation in the WSIS, to the point where, in certain cases, it is likely that some of them participated in the WSIS explicitly because of its linkages to ICANN. Thus, it would be overly simplistic to suggest that the lesson here is that ICANN emerged as a logical follow-up venue and that some organized sub-component of WSIS CS migrated there *en masse*. It would be equally misguided, however, to neglect what the NCUC episode suggests about how the WSIS experience has transformed civil society participation.

The supporters of the NCUC were able to organize and mount sustained resistance to what they perceived to be efforts on the part of ICANN staff to impose a civil society structure distinct from their own model. These efforts included a highly successful membership drive as well as a campaign to mobilize the voices of civil society actors not typically involved in ICANN in order to register a general sense of "disbelief" and "injustice" about the non-responsiveness of ICANN staff to consensus-based, bottom-up policy development.[15] Each of these drew on networks, relationships and communication platforms that had been developed during the WSIS and IGF processes. The ultimate aim was to demonstrate the viability of the proposed NCUC structure, its conformity with proposed rules and its broad support within global civil society circles to the ICANN board of directors in the hope that they would set aside the separate plans that policy staff had been pushing.

It is impossible to know how much its WSIS experience contributed to the ability of the NCUC to successfully navigate the politics associated with

---

12 See the Non-Commercial Users Constituency (NCUC) website. http://ncdnhc.org/
13 Milton Mueller, *Civil Society at ICANN: A Success*. [Governance]. (November 1, 2009).
14 For example: Robin Gross, Milton Mueller, Avri Doria, Bill Drake, Carlos Afonso etc. Full lists of membership and officers can be found on the NCUC website referenced above.
15 See Milton Mueller, *Is ICANN Listening?* [Governance]. (July 23, 2009).

institutional bureaucracies and decision-making. It is certain, however, that the ability to effectively organize a constituency and develop internal structures that conform to institutional guidelines contributed fundamentally to the success of the NCUC proposal, as did the ability to mobilize the support of wider civil society at times of tension and crisis. The ability of NCUC leadership to accomplish each of these elements was certainly augmented due to their participation as individuals in the WSIS. Generally, the NCUC episode illustrates the extent to which the capacity of individual civil society actors to participate effectively and strategically in institutional politics developed and expanded over the course of the WSIS and the extent to which the inter-connections between civil society actors that were established in the form of semi-formal networks and communities of practice have transformed the ability of civil society actors to meaningfully impact global governance in the post-WSIS environment.

OECD

In June 2008, a wide swath of Internet governance stakeholders from government, civil society and the private sector were brought together around an OECD ministerial meeting in Seoul, Korea, on "The Future of the Internet Economy." A background report entitled "Shaping Policies for the Future of the Internet Economy" recognizes that "the open and collaborative nature of the Internet challenges traditional policymaking." In response it suggests that

> a multi-stakeholder approach to achieving an appropriate balance of laws, policies, self-regulation and consumer empowerment may be the only way to promote the Internet economy effectively. An effective and innovative multi-stakeholder approach has to be developed for government, the private sector, the technical community, civil society and individual users to join forces in shaping the policy environment for the future of the Internet economy.[16]

In the OECD's view, more effective Internet policy-making is directly linked with improvements in our ability "to boost economic performance and social well-being, and to strengthen societies' capacity to improve the quality of life for citizens worldwide."[17]

---

16   OECD, *Shaping Policies for the Future of the Internet Economy*. http://www.oecd.org/site/0,3407,en_21571361_38415463_1_1_1_1_1,00.html
17   Ibid.

Despite the fact that the OECD has well established structures for facilitating private sector and trade union participation in its processes,[18] accepting the orthodoxy that global Internet governance must involve some form of multi-stakeholder model was not an intuitive move for the OECD. A prominent figure in WSIS civil society reported that OECD staffers had informally responded to his queries about the status of civil society at the Seoul meeting by saying that the OECD was concerned that whatever mechanism for CS participation they might establish would determine whether or not the people and organizations presenting themselves as CS would be acceptable and accountable to other CS actors.[19] Regardless, the OECD has set up formal structures to engage civil society through the creation of a Civil Society Information Society Advisory Council (CSISAC).

WIPO

In her "institutional overview" of the World Intellectual Property Organization (WIPO) for the 2007 Global Information Society Watch, Robin Gross[20] laments that "While WIPO boasts that over 250 NGOs and IGOs currently have official observer status at WIPO, the vast majority of these NGOs are trade industry organizations from wealthy countries participating for the purpose of maximizing private gain."[21] Gross also points out that the participation of civil society in WIPO has traditionally functioned to stack the deck even further against developing country interests, and that WIPO decision making generally is often seen to lack transparency, undermine democracy and favour the private interest at the expense of the public interest. Gross's article was written as plans were being made for a third round of discussions on a development agenda for WIPO. Given that the initial rounds of discussion over the development agenda had ended in stalemate when faced with objections from certain developed country governments, Gross was, at the time, less enthusiastic about the prospects for round three: "Without support from the wealthy member states, reform at WIPO is almost impossible."[22]

---

18 The Business and Industry Advisory Committee to the OECD (BIAC) and The Trade Union Advisory Committee to the OECD (TUAC), respectively.
19 See Bill Drake, *WSIS Principles and Conferences Date*. [governance]. (March 17, 2006).
20 Gross is a lawyer and the founder and Executive Director of IP Justice, an international civil liberties NGO, and was a prominent figure in WSIS CS. See the IP Justice website. http://ipjustice.org/wp/about/people/robin-d-gross/
21 Robin Gross, "WIPO" In *Global Information Society Watch 2007*. (2007 at p. 65). http://www.giswatch.org/gisw2007/download
22 Ibid. (at 69).

Yet, over the course of 2007 as the WIPO development agenda discussions concluded, James Love, executive director of the NGO Knowledge Ecology International and also a prominent figure in various CS networks, noted that "agreement on dozens of WIPO reforms was broader and more substantive than had been anticipated." With the result, he continues, that "some of the measures signal important changes in this controversial UN body." Civil society was perceived to have played a significant role in this position shift as:

> Many non-government organizations (NGOs) and experts have labored long and hard on the development agenda negotiations...The contributions of the (north and south, east and west) development, consumer, free software, library and public interest groups working on technology issues were very important.[23]

But, as suggested by Gross, something of a cultural shift at the WIPO toward the perspective and input of public interest civil society groups was a necessary precondition for such influence. In his assessment of the proceedings, Love points out that a key element of compromise was the openness of many of the same European states that were engaged with civil society during the WSIS such as Germany, Switzerland and the UK to developing country perspectives. Perhaps civil society was more effective at advocating for the developing country perspective at WIPO as a result of the WSIS experience. Certainly, the involvement of individuals and groups in the WIPO proceedings who had experienced the integration of civil society groups into the WSIS must be seen to have laid the groundwork for this cultural shift. Lessons were undoubtedly learned from the WSIS/IGF experience by CS activists at WIPO, and also by government delegates and diplomats. One observer noted in particular how "the Chair, Ambassador Trevor Clarke of Barbados, steered the helm of the Development Agenda process with judicious authority rejuvenating hopes that WIPO can mainstream public interest concerns into its core mandate."[24] Clarke was involved in the WSIS in many capacities, including as a member of the WGIG.[25]

---

23 James Love, "WIPO Embraces Reform on Intellectual Property Mission." *The Huffington Post*. (February 23, 2007). http://www.huffingtonpost.com/james-love/wipo-embraces-reform-on-i_b_41951.html
24 Ibid.
25 Gwen Hinze, *Blogging WIPO: Progress at WIPO Sets Stage for Second Round of Development Discussions in June*. (February 24, 2007). http://www.eff.org/deeplinks/2007/02/blogging-wipo-progress-wipo-sets-stage-second-round-development-discussions-june

GAID

The UN ICT task force had operated in parallel and as an occasionally overlapping process during the WSIS. A group of CS participants in the WSIS also sat on the task force and contributed to its work. At the conclusion of the WSIS, the task force was effectively re-launched as the Global Alliance for ICT and Development (GAID) as a response to

> the need and demand for an inclusive global forum and platform for cross-sectoral policy dialogue on the use of ICT for enhancing the achievement of internationally agreed development goals, notably reduction of poverty.[26]

The GAID is a self-described "think-tank" largely responsible for awareness raising, cooperation building and advising the Secretary-General of the UN. It is framed as a multi-stakeholder body: its Steering Committee and High-Level Strategy Council each feature a significant contingent of individuals from CS organizations and its structure includes a decidedly CS-focused "Champions Network," defined as

> a group of activists, experts and practitioners promoting development through the use of ICT. They echo and amplify at local, national and regional levels, the lessons learned and best practices identified through the work of the Alliance.[27]

CS participants in the GAID have expressed repeated frustration with this structure, going as far as drafting a letter outlining the need for reform.[28] However, as this book was going to press, the GAID was thought to be on the verge of structural reforms that would produce even less desirable outcomes for civil society. GAID Strategy Council member, Executive Director of the NGO IT for Change and prominent WS CS participant Parminder Jeet Singh reported in January 2010 that the GAID

> may be headed towards getting folded up into a regular UN department, doing mundane work (that's what I fear). [...] What we need instead is a set of more focused and clearer objectives and work plans, and a better network structure focused on public interest actors, chiefly those involved with development issues.[29]

---

26 Global Alliance for ICT and Development website. http://www.un-gaid.org/About/OurMission/tabid/893/language/en-US/Default.aspx
27 For descriptions and membership rosters of each of these internal GAID structures see the group's website (referenced above).
28 See Michael Gurstein, *FW: GAID. RE: [governance] AW: [tt-group].* (December 28, 2009).
29 Parminder Jeet Singh, *IGF and GAID. Re: [gaid-discuss] [governance].* (January 4, 2010).

Thus, despite the GAID initiative being planned explicitly around the involvement of CS, this sort of multi-stakeholder body has proven difficult to administer and produce results within the UN system and the inclination of its organizers is clearly toward further institutionalization as a response.

ITU

The role of civil society participation in the ITU continues to be controversial. In one sense, the ITU can claim to be more open to stakeholder participation than many other UN organizations. The ITU has been granting formal status to non-governmental stakeholders since the early part of the twentieth century. Business entities (and also certain larger not for profits and NGOs) cooperate with ITU through so-called "sector membership."

But, sector membership is very different from the way NGOs participate in other UN Agencies and Programs in that ITU sector members pay a fee (which can be waived under certain circumstances and often is for not-for-profit entities). Sector members can participate in ITU working groups and therefore are positioned to influence decision-making from the ground floor of the ITU's policy cycle. The largely technical decisions taken by the ITU impact the telecom industry and sector members are willing to pay for their participation precisely because of the influence over ITU decision making that sector membership provides.

In contrast, the classical "observer" status for NGOs and civil society that has existed since the creation of the UN is usually limited at ITU to plenary and subcommittee meetings. NGOs are permitted to submit positions in writing and, in certain cases, to make plenary interventions. But they are not involved in "negotiating" per se in that observers usually cannot participate in closed meetings and working groups and are not considered to have any sort of voting interest in the decisions that are taken by the intergovernmental process.

Thus, the question of reforming civil society participation in the ITU is complex. Though the "observer" status common to many UN agencies and programs does not exist at present in the ITU, a case can be made that the current "sector member" status actually allows for a greater degree of participation. This is of course problematic because of the restricted membership and fees, but, from this perspective, the campaign to push the ITU to create the "observer" status that civil society and NGOs enjoy elsewhere in the UN system could also enable the ITU to roll back the participation rights of civil society. The bottom line, however, is that "sector membership" is not an appropriate role for civil society at the ITU.

In an email sent to a civil society listserv, WSIS executive director Charles Geiger drew on his experience working with both the ITU and civil society over the course of the WSIS to reflect on this question. "The 'sector member' status does not exactly fit for civil society entities that do defend general societal interest like Human Rights, Access to Knowledge, ICTs for Development etc." Geiger continues that "such NGOs"

> are used to the "observer" status in other UN entities, which is free of cost, and do not see any interest in paying a fee for becoming ITU sector members. They do not want to participate in working groups, they want to speak out in Plenary meetings. They consider their participation as political, not technical...In my view, ITU is not reluctant to deal with civil society, the problem is different: As a technical organization, ITU never felt the need to create an observer status for political participation of civil society. But... Internet Governance is a highly political theme, and if ITU wants to play a role in this field, it will have to open up to civil society and to create a format for meaningful participation of civil society representatives.[30]

Geiger's commentary on the tension between ITU sector membership and observer status brings into relief some interesting observations about the place of civil society within the global governance of communication. International organizations are increasingly recognizing that they have to include civil society, somehow, in their deliberative processes for a variety of reasons: to gain legitimacy, benefit from expertise, and catalyze community involvement. The future of more political stakeholder participation in the ITU is being studied and may ultimately prove to be a tipping point in the emerging regime of global Internet governance.

Only days after the first IGF in Athens, the ITU held its 2006 Plenipotentiary Conference in Antalya, Turkey. One of the outcomes of this meeting was the adoption of Resolution 141 (Antalya, 2006) entitled "Study on the participation of all relevant stakeholders in the activities of the Union related to the World Summit on the Information Society." Resolution 141 instructed the ITU Council to establish a working group to study the issue and prepare, by the end of 2007, a report on existing mechanisms within the United Nations, other UN specialized agencies, and intergovernmental organizations.

By way of follow-up, the Antalya Plenipotentiary Conference resolved to launch the Fourth World Telecommunication Policy Forum (WTPF), "in order to discuss and exchange views regarding Internet-related public policy matters, among other themes." The role of the WTPF is to "prepare reports and,

---

30 Charles Geiger, *RE: ITU and ICANN—A Loveless Forced Marriage*. [WSIS CS-Plenary]. (December 3, 2008).

where appropriate, opinions" for further consideration of possible reforms to the International Telecommunication Regulations at a World Conference on International Telecommunications (WCIT) to be convened by ITU in 2012. The WTPF forum was planned for Lisbon in April 2009.[31]

A so-called Informal Consultation Meeting between CS and the ITU was organized to contribute to the tasks of the ITU working group created by Resolution 141. The meeting was attended by prominent WSIS CS actors including representatives from CONGO and the APC. Former WGIG member Bill Drake was a driving force behind the meeting and gave an extensive presentation. The role of CS within the ITU was compared unfavorably to other intergovernmental organizations. A variety of recommendations were made that aimed at making the ITU more open to both the observer status available to CS elsewhere within the UN system, and the multi-stakeholder partner status emerging in the IGF and elsewhere.[32]

With its working group study on the role of civil society still in progress, the ITU determined that individual members of the public may fill out a form attesting to their "proven interest in matters related to the WTPF-09, along with expertise and experience in Information Society issues," with ITU staff then deciding if they qualify to attend, but not directly participate in the meeting.

These cautious steps toward the creation of some form of stakeholder observer status can only be seen to represent a serious step back from the participation rights that civil society procured from the ITU-hosted WSIS. Whether functioning as host of a World Summit or organizing its internal policy development plans, the ITU, as an intergovernmental organization, must be responsive to the views of its membership on this issue. This can, as it was at the WSIS, be a blessing for civil society's ambitions. Charles Geiger is clear to point out that "the way civil society was handled in WSIS was mostly decided by the WSIS Intergovernmental Bureau, where the decisive influence did not come from ITU, but from the two PrepCom presidents, Adama Samassékou and Janis Karklins."[33] But the extent to which the organizational policies of the ITU must reflect the views of its membership seems more likely, in the long term, to present a roadblock to civil society.

---

31 ITU, *Final Acts of the Plenipotentiary Conference (Antalya, 2006): A Selection of Internet Related Resolutions*. (November 2006). http://www.itu.int/osg/csd/mina/index.html
32 For details on the event including the submissions and presentations made in its final report, see the WSIS official website. http://www.itu.int/wsis/implementation/2007/civilsocietyconsultation/index.html
33 Charles Geiger, *RE: ITU and ICANN–A Loveless Forced Marriage*. [WSIS CS-Plenary]. (December 3, 2008).

Thus, the real solution may lie in a continued commitment on the part of CS to remain mobilized around the issue of multi-stakeholder participation, using arguments about legitimacy and transparency to influence various international organizations until the culture of multi-stakeholderism in global governance is accepted to such a degree that certain governments are unable to impede it in organizations such as the ITU (or, at least, unwilling to expend the political capital that would be required to do so). In the concluding chapter of this book, we reflect on the legacy of the WSIS and its role in the realization of such ambitions.

• CHAPTER NINE •

Multi-stakeholder Global Governance at the WSIS and Beyond

What is the legacy of the WSIS? Does the WSIS represent a rebranding of existing practices, or a new politics? Will the WSIS merit more than an asterisk in the history books? Did the WSIS bring about a cultural change in the way people see democratic global governance? What has changed as a result of civil society involvement in the WSIS? In this chapter we reflect on the experience of civil society at the WSIS and since in the effort to present some answers to these questions.

The Substantive Legacy of the WSIS

In terms of substance, the WSIS produced three principal outcomes: the WSIS principles for the information society, accompanied by an implementation plan; a preliminary effort to respond to the digital divide; and the incubation of a global governance approach to Internet governance.

The WSIS was a point of convergence where activists, researchers, international civil servants and representatives of business and civil society organizations came together to confront issues of development and social justice through the lens of communication. As previous UN Summits had done for the environment (Rio), for women (Beijing) and for a host of other themes, the WSIS helped crystallize an approach to a set of critical issues facing society and above all a set of new perspectives on those issues. In this respect, the accomplishments of the WSIS were threefold.

First, the WSIS achieved formal recognition amongst the international community of the close ties between information, communication and development. In and of itself, this step is significant and highlights the pedagogical function realized by the summit. Following the Millennium Development Goals' emphasis on the need to make available the benefits of ICTs for development (Goal 8, target 18), the WSIS highlighted many of the social, economic, cultural and democratic gains that can be associated with the widespread diffusion and use of these new technologies (see the Geneva Declaration of Principles). It also defined a path for the international community to follow (see the 11 action lines of the Geneva Plan of Action) and a series of precise objectives (presented in the Geneva Plan of Action and the Tunis Agenda for the Information Society). Thus the WSIS explored, clarified and articulated a view on the complex relationship between technological innovation, information, culture, knowledge, communication, and international development.

By virtue of both its failures and its successes, the WSIS positioned global communication issues as complex and multidimensional problems that have more to do with structural inequalities and injustices than with some intrinsic characteristic of technology. This view is magnitudes more sophisticated than the sort of clichés about the eclipse of the industrial age by an information revolution that typified much political discourse around the information society prior to the WSIS.

Finally, by linking new information and communication technologies (ICTs) with the challenges of international development, the WSIS contributed to further establishing communication as a cross-cutting issue for global governance and to bringing together representatives of various established local, national and global movements and organizations around it.

As such, the World Summit on the Information Society helped bring consistency and unity to multiple issues associated with access, control and use of knowledge, culture, and information. The summit repositioned themes that were previously largely seen as secondary or technical issues in a larger struggle for social justice and framed them as part of the foundation of modern society—making clear, in the process, that this foundation is currently built upon exclusion, inequality and injustice. Present every step of the way, civil society actors consequently framed the rapid technological changes under way as opportunities to fight persisting social and economic inequalities. Overcoming the digital divide—an objective repeated ad nauseam in the official documents of the summit—was understood to be instrumental to the bridging of the many social, economic, and political divides that persist in the modern world. The WSIS was thus part of an international trend toward wider recognition of the fundamentally political nature of technological development, media governance and the distribution of communication resources amongst different social and economic actors. The WSIS Declaration of Principles produced at the conclusion of the Geneva phase stands as the most developed intergovernmental normative consensus on global governance goals for the information society.

The political vision developed in this document has more than a simple declaratory value; it is also an analytical and conceptual tool helping the international community understand a set of nebulous concepts—the notion of a global information society the most prominent among them—which are then linked to specific themes, issues and challenges. The Geneva Declaration of Principles thus has to be understood as a reference point for addressing a wide range of international development issues from the perspective of information and communication. The Geneva Plan of Action, as well as the subsequent work accomplished in Tunis leading to the adoption of the Tunis Commitment and Tunis Agenda for the Information Society, aimed at translating this political vision into concrete action in order to enable its implementation. However, as we saw earlier, most assessments of the political process are that it was, at best, disappointing.

But the political failures of the WSIS were offset by significant cultural gains. When approached not just as a political summit, but also as a site from which the production and dissemination of critical discourses on communication emanated, the WSIS acquires a whole new degree of relevance. The WSIS conferred legitimacy, coherence and unity onto these issues. For the first time, a UN conference at the highest level gathered the four major categories of stakeholders—governments, international organizations, the private sector and civil society—around key issues of communication. More fundamentally perhaps, this summit contributed to the establishment of working relationships between many of the actors involved directly or indirectly in its activities. These networks established mutual understanding and solidarity between actors at the event, and have facilitated the dissemination of the knowledge acquired at the summit in other settings. The Civil Society Declaration, *Shaping Information Societies for Human Needs*, produced at the end of Phase I of the WSIS, and the CS statement *Much More Could Have Been Achieved*, issued at the end of Phase II, stand as landmark documents in this regard.[1]

The gap between recognition and action is evident elsewhere within the so-called WSIS principles as well. For example, the need to address the digital divide was essentially agenda item one for the WSIS. It was arguably the most logical issue for the WSIS to address from the start as well as the topic which all actors agreed had to be included in the WSIS discussions. But recognition at the WSIS of the existence of the digital divide was not followed by a consensus on action. In this sense, the outcome of the WSIS reinforced the extent

---

[1] See appendix of this volume. Full citation: WSIS Civil Society Plenary, *Much More Could Have Been Achieved: Civil Society Statement on the World Summit on the Information Society*. (December 18, 2005 Revision 1—December 23, 2005). http://www.worldSummit2003.de/download_en/WSIS-CS-Summit-statement-rev1-23-12-2005-en.pdf

to which there is still a rich-poor, North-South power divide in the world and underscored the fundamental resistance of those who have wealth to share it equitably with those who do not.

The issue of Internet governance presented itself in a different manner. In many respects, the WSIS sanctified recognition of IG as a field of international public policy. Before the WSIS there was debate over whether Internet governance of some form was desirable or even possible. Important stakeholders and respected analysts alike were saying that the Internet could not be regulated, that the technology could not be controlled in the way that public authorities had enforced constraints on the use of other platforms of mass communication. Other specialists argued that the key to the Internet lay in its capacity to facilitate its own innovation and that, by the early 2000s, development of the Internet was only in an embryonic state. The WSIS arrived, acording to this view, too soon to know where the development of the Internet was headed and too early to start taking steps that might interfere with its innovation through the imposition of governance structures.

The discussions that took place at the WSIS, however, led to broad acceptance that the Internet not only could be but was being controlled and that the public policy issues it raised required some form of institutional governance. In the process, discourse on Internet governance shifted away from "can the Internet be controlled?" toward "who does and should govern the Internet?, how is the Internet being governed? and how ought this system to function?" WSIS, in effect, legitimized Internet governance as an area of public policy concern, and, however timidly, took some small steps toward dealing with it. The WSIS not only proposed a multi-stakeholder-driven definition of Internet governance, but also indicated a set of political parameters, potential participants and an institutional framework for dealing with these issues.

Over the first phase of the WSIS in particular, a variety of constituencies mobilized and struggled to get language into the official documents around gender, youth, disability and many other issues. Arguably, the multi-stakeholder governance model is the WSIS principle that has been most effectively claimed elsewhere since the WSIS. The Geneva declaration (at para 17) reads:

> **We recognize** that building an inclusive Information Society requires new forms of solidarity, partnership and cooperation among governments and other stakeholders, i.e. the private sector, civil society and international organizations.

In this sense, the legacy of the WSIS principles, and indeed of the entire WSIS itself, arguably hinges on its processional and institutional innovations. Yet, it is in part precisely through the substantive contributions that civil society has made by leveraging the multi-stakeholder practices endorsed by WSIS that these processional elements have resonated.

Civil Society and Global Governance of Communication

Official outcomes such as the Geneva Declaration of Principles were not the only products of the WSIS process. While the WSIS official documents show a snapshot of the intergovernmental consensus at a given time—the lowest common denominator that could be agreed upon by the delegations in the room—the civil society documents and perspectives articulate a visionary perspective on how institutions of global governance could deploy mass communication to the realization of social justice, human rights and development.

A fundamental question that emerges from the WSIS experience is this: did the inclusion of civil society in the policy development process contribute to the incubation of a new social movement around global communication governance?

Social movements represent an attempt to recode social reality, to problematize assumptions that had previously gone unquestioned, to critique conventionally accepted definitions of what constitutes the good, the just and the doable for society and for public authorities. The notion of a social movement refers to a process by which large numbers of people get to, in effect, reframe their world and act on such understandings in the effort to incite social change from the bottom up.

In part, the experience of civil society at the WSIS ran exactly counter to this normative view of a social movement, in the sense that the UN tried to push an agenda top-down and convince participants that issues related to the "information society" should matter to them in a fundamental way. Not all of civil society accepted the premise that participation in a World Summit on the Information Society was important. In other words, the central problematic of the WSIS was not defined by a rising public consciousness manifesting itself through public opinion. The potential for grassroots mobilization on issues discussed at the WSIS was always contingent upon the ability of enlightened individuals, experts and representatives of grassroots organizations such as the Association for Progressive Communications (APC) and the World Association of Community Radio Broadcasters (AMARC) to connect people with issues and knowledge.

At the same time, a series of grassroots groups came together around global communication governance issues in an unprecedented manner. The last time such questions had attracted significant intergovernmental policy attention, during the UNESCO MacBride Commission debates of the early

1980s, it had been largely an affair of states. The openness of the WSIS process to civil society participation not only encouraged the founding of more such groups, thus broadening the institutional experience and trajectories of the individuals who were involved, it has greatly enhanced the number of civil society activists with policy knowledge, experience and interest in working on issues of global communication governance. This may only mean an incremental change in the process of democratization of communication, but it is light years ahead of what was barely imaginable in the way of global communication governance activism before. The WSIS created a cohort of several hundred people empowered with the knowledge of how communication politics and governance processes function who are now sharing that experience with their networks and applying it in their everyday activities.

Furthermore, the WSIS was an important moment where issues surrounding communication and technologies were recognized by various groups and social movements as meta-issues impacting social justice and democratization. Communication is a meta-issue demanding activist attention in the sense that social movements increasingly depend on functional systems of mass communication and, crucially, on having their rights to free speech and public assembly protected within such communication systems. Typically, however, social movements tend to focus on specific vertical issues (employment, gender, development, etc.). Although communication cuts across these issues horizontally, it tends to receive scant attention in broadly based campaigns for social change and social justice.

Civil society participation in the first phase of the WSIS seemed to be making inroads into raising the sort of awareness that would be required in order to mobilize a broad-based social movement around global communication governance and to create the large global network of activist groups that would be required to sustain it. But, over the course of Phase II, this momentum dissipated. The WSIS CS space was quickly populated by groups more interested in specific communication policy issues rather than general social justice concerns. Some of this was due to the agenda changes that were determined by the WSIS itself. Some of it was related to issues of burnout or a lack of sustainability, as individuals and groups who had been at the centre of civil society participation in Phase I lost the interest, funding, time or all of the above that would have been required to stay intensely involved over the course of two additional years of meetings and negotiations.

In this sense, mobilization around global communication, in spite of the participation of civil society in the WSIS, is still below the threshold of a social movement. For example, the post-WSIS experience of the CRIS campaign and others suggests that it remains a challenge to get the World Social Forum to even recognize these issues. Going forward, civil society needs to ask itself why it is so hard to mobilize at the grassroots level around global communication governance issues and to meaningfully frame them as intimately linked to more general movements for social justice. More needs to be done to bring mass communication closer to people, but a necessary first step in this mobilization seems to be to raise awareness that such issues fundamentally matter.

Does this mean that the WSIS efforts to include CS in global governance, and the entire WSIS process alongside it, should be judged to have failed to contribute substantially to the democratization of global communication? There is an argument to be made that social movements do not fail, that their very existence assures some measure of social change, no matter how slight and regardless of whether or not their vision of a new society is eventually fully realized. It is too early to make final evaluations of what civil society accomplished at the WSIS, but it seems that the "do not fail" principle applies equally here, whether the conclusion is that there is an embryonic social movement emerging or nothing of this sort. On the basis of the WSIS experience it is clear that it is better for activists to be inside the processes of global communication than to be excluded from them. It is of course equally evident that civil society should work on multiple levels at once: at global, national and local venues, from inside the official policy process and outside.

Whatever else can be said about the WSIS, there has never been a bigger mobilization of communication activists and global communication governance has never been as much of a focus for global activism. An embryonic movement is taking shape, akin to the environmental movement of the early 1970s. It was a good decade after Rachel Carson published *Silent Spring* in 1962[2] before a plethora of disparate and mostly marginal groups crystallized into a worldwide social movement around environmental protection.

A relatively small number of groups have been working on global communication governance issues for a long time. In addition, there are vibrant media reform movements emerging in various countries, the US most remarkably. Prior to the WSIS, however, there was little transnational media activism nor consciousness that media issues needed to be addressed beyond the national level. Through civil society participation in the WSIS, awareness, knowledge and experience about the importance of global governance of communication has been gained within civil society. Even during the second phase when a relatively more narrowly focused group of often individual delegates emerged, connections were made, sustainable, professionalized, broad-based grassroots groups such as APC and IT for Change were still around and

---

2   Rachel Carson, *Silent Spring*. New York: Houghton Mifflin. 1962.

civil society people from various constituencies that may not have been at the centre of the WSIS process were still there representing their constituency and bringing new knowledge and experience back to their base. Such developments may eventually trickle down and lead to greater awareness amongst various social movements about global communication governance issues and instigate eventual efforts at coordinated action. They may not. Only time will tell, and it is simply too early to conclude whether or not the WSIS experience was the "Silent Spring" moment of a yet to fully emerge social movement, an important precursor step providing the interconnections that will be required of a mobilization around a yet to occur catalytic moment or simply a point of intersection where a number of disparate ships passed in the night before resuming their separate journeys.

A Constituency in Search of Legitimacy

The participation of non-governmental actors in global political processes built on principles of state sovereignty and intergovernmental negotiation inevitably raises questions of legitimacy. Non-governmental stakeholders have neither political legitimacy (conferred by the control of state apparatuses), democratic legitimacy (conferred by electoral results), nor the economic power to justify, a priori, their presence in UN forums. The active participation of civil society in these arenas is typically justified by claims that their presence provides a sense of moral conscience and transparency, by the social, cultural and political expertise of their memberships, by their proximity to the social groups and/or regions being debated as the object of public policy, by the element of direct democracy their participation provides to the UN process, or by a combination of some or all of the above. In the absence of a clear basis on which to establish rules that legitimized their participation in the WSIS, CS organizations participating in global governance structures tended to defend their involvement with rhetoric centered on arguments for transparency, inclusion, accountability, and political and thematic relevance.

In order to be effective, the discourses articulating these four elements had to apply to the structures and modalities of internal participation used by civil society at the summit. The presence of persistent inequalities in the quality and intensity of participation experienced by different sectors of WSIS CS raised an unavoidable issue of legitimacy. The effort conducted over the course of Phase II to develop more transparent, inclusive, accountable, productive and politically effective participatory mechanisms represented, in part, an attempt to respond to criticism about the legitimacy of interventions claiming to be made on behalf of civil society as a constituency.

These same criteria were employed to counter criticism generated externally about the legitimacy of civil society's activities at the WSIS. Civil society's message over the course of the two phases of the summit was that its presence made the process more transparent and inclusive and contributed to holding governmental delegations accountable to global public opinion. Thus, civil society argued that its participation added value to the WSIS because it conferred legitimacy to the summit, and insisted that civil society was engaged in a good-faith effort to work constructively and effectively with its governmental, institutional and private sector partners. The post-WSIS enthusiasm for multi-stakeholder global governance seems to suggest that this was an effective way of framing civil society participation and of convincing other stakeholders of its contribution.

However, the definition of civil society is fluid and its boundaries are porous, a reality that, on more than one occasion, rendered the participation of non-governmental actors at the WSIS problematic. Difficulties arose over the categorization of certain actors who were affiliated with multiple constituencies. The report of the high-level UN panel on relations with civil society (chaired by former Brazilian president Fernando Henrique Cardoso) defined the concept of civil society in the following terms:

> the associations of citizens (outside their families, friends and businesses) entered into voluntarily to advance their interests, ideas and ideologies. The term does not include profit-making activity (the private sector) or governing (the public sector). Of particular relevance to the United Nations are mass organizations (such as organizations of peasants, women or retired people), trade unions, professional associations, social movements, indigenous people's organizations, religious and spiritual organizations, academe and public benefit non-governmental organizations.[3]

Although conceptually useful, this definition does not take into account the complexities of global politics. For example, the private sector finances non-profit organizations whose primary mandate is to promote, represent and defend their economic interests with political representatives. These lobbying organizations occupy a space at the border of the private and voluntary sectors and may pose significant problems of categorization. Yet, their integration in international forums as civil society organizations leads to a double representation of the corporations or economic sectors whose interests these organizations defend. This integration also extends the legitimacy of private sector

---

3   See Fernando Henrique Cardoso et al., *We the Peoples: Civil Society, the United Nations and Global Governance. Report of the Panel of Eminent Persons on United Nations–Civil Society Relations* (A/58/817). (June 11, 2004). p. 13. http://www.un.org/french/ga/ search/view_doc. asp?symbol=A/58/817&referer=http://www.un.org/french/reform/panel.html&Lang=E

positions by conflating them with those of a constituency that is expected to intervene in the name of the citizenry in international forums. Finally, the presence of such organizations within civil society may complicate the production of discourse that is critical of the private sector by civil society. The integration of organizations maintaining close ties with business interests within the civil society sector participating at the WSIS generated debates and controversies which functioned to illustrate the risks associated with broad and open definitions of civil society.

The boundaries of civil society are also porous when it comes to political influence. The organization of the second phase of WSIS was marked by attempts to co-opt and hijack civil society participative structures by agitators determined to prevent the drafting and presentation of any statements critical of the Tunisian government. Their infiltration of civil society—by entirely legitimate means (according to the established rules)—raised important boundary issues. Some members of civil society also maintained close relationships with their national delegations and were even accredited to the Geneva and Tunis Summits as members of government delegations. These CS participants used the privileges conferred on them by their status as government delegates to enhance their participation in the event. Although this closeness with the delegations enhanced the participation of members of civil society at the WSIS, it may eventually lead to conflicts of interest and raises questions about the independence and impartiality of groups and individuals participating in international forums where governments negotiate policies.

Representation and Performance

The participation of civil society at the WSIS was also marked by issues of representation. Although civil society cannot justify its participation in an international summit by claiming a representative function, WSIS CS was careful to select spokespeople representing various regions of the world (often focusing on the developing world), and from different social groups generally marginalized within global governance. This selection process was used by civil society to mount a public performance aimed at illustrating the diversity of the constituency and strengthening its moral authority by implying that it held a degree of political representivity.

The will to display and publicly promote diversity in the ranks of civil society did not prevent the spread of internal criticism on the issue. There were strong critiques from within CS of its reliance on English as a working language outside the CS plenary and content and themes meetings where UN translators were supplied, of the overrepresentation and prominence of individual and group CS participants based in OECD countries, and of the general lack of ethnic, cultural and geographical diversity amongst CS as a whole. Significant problems limited the participation of CS from developing countries throughout the two phases. These were in large part caused by the exorbitant cost of attending meetings which were, for the most part, held in Geneva, a notoriously expensive destination. Overall, and notwithstanding the efforts that were made to diversify its ranks, WSIS CS could hardly claim to have constituted a representative sample of international civil society.

Such questions about the significance of who did not participate in the WSIS are fundamental to evaluating the role of CS in global governance. But, so are questions about the significance of who did represent CS at the WSIS. The WSIS experience could be seen to have affirmed the role of a highly networked transnational social and political elite at the expense of true grassroots social movements. Participation in the WSIS demanded possession of or access to the capital, knowledge and institutional resources required for active participation in international forums. As a result, many of the prominent CS delegates to the WSIS were academics, professional activists and/or individuals with vast previous experience in the UN system, adept at navigating through the topics being discussed, taking advantage of the opportunities presented to them (including for personal professional advancement) and at securing available funding. This does not mean to imply that there was no grassroots involvement at the WSIS, rather that the constitution of WSIS CS reflected the inaccessibility of a global governance system defined by the complexity of UN processes and procedures, a professional culture that favours established players and highly educated individuals, and the prohibitively high costs of attending meetings.

Strategy and Effectiveness

The integration of CS into an intergovernmental policy forum such as the WSIS raised difficult questions in regard to effective participation. Two key issues stand out above the rest. First, did the participation of these groups contribute to enriching the political discussions, to promoting dialogue and broadening the spectrum of perspectives, issues and positions presented at the WSIS? Second, would civil society have been better off mobilizing outside of the formal structures of the summit in order to avoid the possible political cooptation and dilution of its positions?

The integration of civil society at WSIS had clear—although in some respects disappointing—political impacts and unquestionably contributed to widening the parameters of the discussions that took place at the summit. On one hand, various groups and civil society organizations worked over the course of the Geneva preparatory phase to refocus an agenda that was, at the time, heading in the direction of technological determinism, utopian discourses on the information society and a profoundly neoliberal approach to international development. These trends were balanced by the intervention of discourse around issues such as sustainable development, social inequality and exclusion. Civil society also played a political watchdog role at the summit, helping to critique and rally support against certain truly draconian initiatives that were proposed, in particular around human rights. The constant vigilance of civil society on human rights issues formed the basis of an ultimately successful push to convince governments to include strong references to and support for the International Bill of Human Rights in WSIS documents. Finally, although civil society was critical of the political results generated by the WSIS, the official documents expressing the political consensus reached in Geneva and Tunis reflected issues that civil society fought hard to get on the WSIS agenda, initiatives originally proposed by civil society and positions defended by civil society over the course of both phases. In other words, the voices of civil society at the WSIS did resonate amongst governmental delegations and the views and ideas of civil society did contribute—to a limited degree—to the outcome documents of both WSIS phases.

However, the absence of more meaningful influence for civil society on official summit documents led some CS actors to question the relevance of their involvement in the formal structures of the WSIS. The return on investment of CS's political engagement in four years of expensive and time-consuming preparatory process does not seem, prima facie, to have been justified. In addition, the risks of cooptation, collusion and institutionalization associated with mobilizing around such an official and political process might be seen to have weakened the moral authority of the CS participants to stand in opposition to the decisions adopted on behalf of the public interest. In other words, it would be logical to conclude that the modest political impact of civil society on the official process did not justify the commitments and compromises that were required. The issue, however, can and should be looked at from another perspective.

Blanket refusals to engage in direct discussion with policymakers and participate in the institutions where policies are made are simply counterproductive. Civil society has a place within the structures of global governance and a role to play in them. The modest political influence realized by WSIS CS reflects a degree of political and institutional resistance to change that was evident over the course of WSIS.

The WSIS illustrates the need for civil society to continue, to repurpose a phrase used previously by one of the authors of this book, its "long march through the institutions"[4] and thus to contribute to an ongoing process of democratization of global governance. The WSIS was the site of a sustained procedural activism organized by the civil society organizations participating at the summit. A considerable share of the energy invested in the WSIS was devoted to enhancing the level of inclusion of civil society groups in working sessions, to increasing the speaking time allotted to non-governmental actors, to expanding access to various intergovernmental groups and to establishing new and better standards for participation and inclusion.

CS at WSIS in Context and Conceptions of Multi-Stakeholder Democracy

The legacy of this summit is thus larger than its political conclusions. The event set the tone for a renegotiation of the role and place of civil society within global governance. Individuals and organizations who participated at the WSIS as civil society did so with the expectation that the gains made in terms of inclusion and participation would contribute to a campaign for the democratization of global governance that was just beginning. The WSIS thus contributed to laying the foundation for a multi-stakeholder model of governance. This is a significant political achievement in and of itself. The WSIS truly was an event situated at the crossroads of divergent conceptions of global governance.

The WSIS occurred in the midst of a parallel UN process convened to reflect on how democracy within global governance might be strengthened through institutional reform. The report of the Cardoso panel, which focused on "civil society access to and participation in UN deliberations and processes"[5] was at the centre of these efforts. Comparing the experience of civil society participation at WSIS to the conclusions of the Cardoso report reveals a great deal about the differences between theory and practice in multi-

---

4   Marc Raboy, "Communication and Globalization: A Challenge for Public Policy." In David Cameron and Janice Gross Stein, (eds.), *Street Protests and Fantasy Parks: Globalization, Culture, and the State*, Vancouver: UBC Press, 2002: 109–140.

5   See Fernando Henrique Cardoso et al. *We the Peoples: Civil Society, the United Nations and Global Governance. Report of the Panel of Eminent Persons on United Nations–Civil Society Relations* (A/58/817). (June 11, 2004). http://www.un.org/french/ga/search/view_doc.asp?symbol=A/58/817&referer=http://www.un.org/french/reform/panel.html&Lang=E

stakeholder global governance as well as about the extent to which the WSIS followed and deviated from the UN road map for reform.

In terms of civil society inclusion in the summit itself, issues underlined by the Cardoso report were manifest during the WSIS, chiefly: the trouble of politicized, bureaucracy-intensive criteria for accrediting civil society organizations and a lack of meaningful funding programs to support the participation of civil society organizations from developing countries.

Much of the discussion in the Cardoso report centres on the viability of establishing new multi-stakeholder forums and integrating them into existing UN practices. The report encourages the UN to "embrace greater flexibility in the design of UN forums" (at sections 43-49) and to "support innovations in global governance" (at sections 50-56). Some of the initiatives proposed seem to be reflected in the meeting formats used in the IGF and the WSIS follow-up activities. These sorts of small scale, multi-stakeholder responses to the emergence of specific issues or controversies are preferable venues for UN activity, according to Cardoso, than larger conferences and World Summits, for which the report suggests "member states seem to have little appetite...seeing them as costly and politically unpredictable" (at para 58). Cardoso does, however, suggest that "the planning of future conferences should allow stronger roles for the major networks of civil society and other constituencies" without going into further details on how this could be accomplished institutionally.

The WSIS obviously represented an attempt to integrate multi-stakeholder governance into the existing model of a largely intergovernmental summit. In this sense, the WSIS seems to have gone beyond the spirit of the Cardoso panel's view that new and innovative parallel multi-stakeholder processes can buttress the legitimacy of and reduce the need for more traditional governance mechanisms like World Summits. The Cardoso report's failure to either go into specific details on how summit planning processes could be reformed to strengthen civil society participation or to include reference to the WSIS experiment of attempting to do so (that was, by the publication of the report, more than 3 years and the entire Geneva phase in process) may underscore a lack of political will for presenting the WSIS as a precedent for future UN activities. Time and the experience of future summits will tell whether this was simply an oversight on the part of the Cardoso panel or an indication that UN enthusiasm for multi-stakeholderism relates more to parallel processes than to high-level, high-profile intergovernmental negotiation forums.

On a separate level, the WSIS experience raises fundamental critiques about some of the conceptual assumptions that the Cardoso report makes about the role of civil society in global governance. The report suggests that "citizens increasingly act politically by participating directly, through civil society mechanisms, in policy debates that particularly interest them." Concluding that "traditional democracy aggregates citizens by communities of neighborhood (their electoral districts), but in participatory democracy citizens aggregate in communities of interest," the Cardoso report seems to suggest that multi-stakeholder global governance constitutes part of "a broadening from representative to participatory democracy." However, the experience of the WSIS suggests that it is at best naïve and at worst misleading to equate civil society participation in global governance with democracy in any normative sense. The questions of legitimacy, transparency and in particular of representation that everyone—civil society actors included—asked of civil society delegates to the WSIS simply proved to be intractable. At the Tunis Summit, Pakistani Ambassador and Chair of the Internet governance negotiation subcommittee Masood Khan was reported to have implied that civil society had represented the global public in the negotiations.[6] However, even civil society delegates were quick to distance themselves from any such pretense.

These comments were pounced on by critics of CS at WSIS, such as Canadian academic, community networking activist and WSIS participant Michael Gurstein. In an article distributed on various WSIS CS listservs and websites, Gurstein problematized the notion of framing a selective collection of activists who are able to present themselves in Geneva for meetings as "the" representatives of the global public and pointed to Khan's comments as evidence that this fraught logic was nonetheless embedded in the WSIS multistakeholder governance principle.[7] In a reply that was circulated through many of the same CS communication channels, APC's Willie Currie responded that Kahn's comments, though obviously naïve (if not bogus), did not in fact reflect how the majority of CS conceived its own role in the WSIS. It was wrong to assume, Currie wrote, that

> civil society activists engaging the WSIS process agreed with Ambassador Khan that they represented everyone else. This was simply not the case, however flattering Ambassador Khan's remarks.[8]

Conceding that there was some merit to Gurstein's critique of WSIS CS participation as elitist, detached from the grassroots and, generally as too focused on creating opportunities for networking, Currie responded with a results-focused assessment that such lines of critique present

---

6   See Michael Gurstein, *Networking the Networked/Closing the Loop: Some Notes on WSIS II.* (December 22, 2005). http://www.worldsummit2003.de/en/web/847.htm
7   Ibid.
8   See Willie Currie, *Creating Spaces for Civil Society in the WSIS: A Reply to Michael Gurstein.* (December 22, 2005). http://www.worldsummit2003.de/en/web/848.htm

too partial a view and dismisses the real gains that have been made by civil society participation. Remove civil society from WSIS and there would be no IGF, no new global policy space for considering broad public policy issues affecting the Internet, including access to the Internet and the digital divide.[9]

This episode underlines a series of tensions in the multi-stakeholder model. If not a nascent forum of participatory democracy, then how should multi-stakeholder governance be conceptualized? Does the inclusion of civil society in intergovernmental politics actually strengthen democracy within global governance? Is formal participation in institutional governance structures the most relevant way for CS to influence the global governance of communication going forward?

Models of Global Governance

In a paper on the global governance of communication published in 2002 (as the Geneva phase of the WSIS was starting to come together), one of the authors of this book presented four models for what he called "the regulation of access to communication" at the global institutional level.[10] These were:

The libertarian model—no regulation. This was the approach taken by most national regulators in regard to the Internet during the 1990s mainly because they did not know what to do or how to do it. It was also the approach adopted by defenders of the status quo during the early stages of the WSIS debates on global Internet governance. It was suggested that the libertarian model was, at the time "also largely favoured by grassroots activists who are benefitting from this open communication system."

Self-regulation: this is the approach most often favoured by industry players, often with the encouragement of national regulators. In 2002, it was frequently being touted as the solution to problems such as abusive content and the protection of rights, on the argument that consumers will respond if they are not satisfied. But, examples from the corporate sector suggested that even in 2002 the promoters of self-regulation were already recognizing the need for a global structural framework for communication activity, within which industry self-regulation would take place.

The closed club, or top-down institutional model: where plans are negotiated in organizations such as the OECD, G8, G20 or WTO, as well as in the

---

9   Ibid.
10 Marc Raboy, "Communication and Globalization: A Challenge for Public Policy." In David Cameron and Janice Gross Stein, (eds.), *Street Protests and FantasyParks: Globalization, Culture, and the State*, Vancouver: UBC Press, 2002: 109–140.

new institutions emerging as the corporate sector fills the vacuum created by the retreat of national governments from regulatory issues.

The long march through the institutions: a process that is tied to the broader project of democratization of global governance, reflected in some of the initiatives around UN reform and again in notions such as "cosmopolitan democracy." Access to global policymaking was already being fostered to some extent by some important initiatives in multilateral discussions which had demonstrated some openness to the concerns of civil society and the inclusion of NGO representation in their activities.

Revisiting the Model in 2010

It is relatively clear that the libertarian view is now essentially dead. Over the course of the WSIS, as issues related to ICANN and international interconnection charges were investigated and debated by the international community, the fundamental question shifted from "can the Internet be regulated?" to "who should regulate the Internet, how and in what interests?" In the process, it was revealed that many of the actors promoting the libertarian view that the Internet cannot and should not be regulated were in fact themselves benefitting from or directly involved in self-regulatory activities. The global financial collapse of 2008–2009 and the role that lax oversight of the banking industry is perceived to have played in enabling the collapse has further discredited the libertarian model and made it—for the time being at least—a political non-starter where other sectors that fundamentally underpin global capitalism, communication included, are concerned.

Over the course of the WSIS, the position of many of the defenders of the status quo in global Internet governance re-framed their positions to reflect less on the libertarian model and more explicitly promote self regulation. Even the private sector expressed a strong preference for a regulatory approach that would be light-touch and hands-off, but transparent and predictable. Thus, the self-regulatory model has arguably become only more formidable since this typology was first developed in 2002.

In addition, the private sector emerged from the WSIS, just as civil society did, more experienced in and organized for participation in global communication governance decision making. Through the structures organized by the International Chamber of Commerce (ICC), including the Coordinating Committee of Business Interlocutors (CCBI) and BASIS (Business Activities to Support the Information Society) initiatives,[11] the capacity of the lobby for self-regulatory

---

11 See the ICC website. http://www.iccwbo.org/basis/id8213/index.html

models of global communication has expanded. Through active participation in the IGF and other organizations and the development of proactive, industry-led responses to global communication governance issues such as the Global Network Initiative,[12] supporters of the self-regulatory model are working to respond to the public interest concerns and fears about market failure that are often used to critique self-regulatory communication governance arrangements.

In the post-WSIS environment, in particular in regard to Internet governance issues, a version of the international club model has proven to be particularly influential. Membership in many of these clubs is exclusively defined. The OECD, G8 and G20 exclude all but the world's most important economic players. In contrast, membership in other clubs such as the Council of Europe is based on geographic or ideological proximity. The growing push to use the international club institutional model reflects the extent to which the Internet governance debate—in particular—is being polarized along economic and geographic lines. The interests of developed, Western countries are being challenged by a coalition of interests from non-Western and developing states. This occurred at the WSIS, but has since continued, as discussed, at the ITU. In the absence of an overarching global institutional framework, the use of the institutional model of exclusive international clubs allows the developed world to leverage its significant capacity advantages to make agenda-setting moves and insulate its interests.

But, even when membership in these clubs is not exclusive, organizations that group together some—but not all—governments are increasingly active in global communication governance. For example, the UN Commission on Science and Technology for Development (CSTD), a subsidiary body of the Economic and Social Council (ECOSOC), was made a focal point in the system-wide follow-up to the WSIS. Rather than a comprehensive multilateral assembly in which all states always participate, the CSTD is an intergovernmental body with a rotating membership on which representatives of 43 governments sit at any given time. It also makes allowances for multi-stakeholder participation on WSIS follow-up. The ICANN GAC as well as the IGF and its dynamic coalitions are similar examples of relatively open international clubs in which certain governments—but crucially not all governments at once—are working together, often alongside other stakeholders.

In light of the experience of civil society at WSIS, it seems evident that the long march through the institutions is the only one of the four models that has any space for ordinary people, for bottom-up influence of the agenda and for the potential to abet the formation of a social movement around global communication governance issues and meaningfully impact decision making

---

12   See the Global Network Initiative website. http://www.globalnetworkinitiative.org/

processes in that domain. In other words, the long march through the institutions remains the last, best hope for meaningful intervention of public interest concerns into global communication governance.

The contribution to be made through this approach by CS to the global governance of communication is perhaps best captured by the term "policy entrepreneur," coined by the political scientist John Kingdon. A policy entrepreneur is someone who can:

> be in or out of government, in elected or appointed positions, in interest groups or research organizations. But, their defying characteristic, much as the case of a business entrepreneur, is their willingness to invest their resources—time, energy, reputation and sometimes money—in the hope of a future return. That return might come to them in the form of policies of which they approve, satisfaction from participation, or even personal aggrandizement in the form of job security and or career promotion.[13]

This is not an insignificant role for CS, as Kingdon's research suggests that "an item's chances for moving up on an agenda are enhanced considerably by the presence of a skillful entrepreneur and damped considerably if no entrepreneur takes on the case."[14] While the economic discourse used to frame Kingdon's notion may cause discomfort to those inclined to see social movements as a counterbalance to the hegemony of modern capitalism, his essential point should not. Social movements prosper when a critical mass of coordinated action is achieved across a multiplicity of approaches that interconnects the efforts of civil society working inside, outside and alongside established institutions, and when alliances are made between policy activists and grassroots practitioners, progressive mainstream practitioners and stakeholders representing all facets of the issue in question. By engaging meaningfully and strategically in the long march through the institutions and building the network of CS organizations engaged in shaping the global governance of communication, the policy activism of civil society—however it is labeled—can function to mobilize social justice and public interest perspectives around political agendas and activities.

Thus, we conclude by suggesting that the realities of civil society participation in global governance may be far from any normative ideals, but like parliamentary democracy at the national level, multi-stakeholder global governance may be the least imperfect model yet developed for making global politics more fair, transparent and legitimate to the concerns of a global public.

---

13   John Kingdon, *Agendas, Alternatives and Public Policies*. New York: Harper-Collins Press, 1984 (at 129).
14   Ibid (at 215).

• APPENDIX •

"Much More Could Have Been Achieved"

Civil Society Statement on the World Summit on the Information Society

18 December 2005

Revision 1–23 December 2005

I. Introduction—Our Perspective After the WSIS Process

The WSIS was an opportunity for a wide range of actors to work together to develop principles and prioritise actions that would lead to democratic, inclusive, participatory and development-oriented information societies at the local, national and international levels; societies in which the ability to access, share and communicate information and knowledge is treated as a public good and takes place in ways that strengthen the rich cultural diversity of our world.

Civil Society entered the Tunis Phase of WSIS with these major goals:

Agreement on financing mechanisms and models that will close the growing gaps in access to information and communication tools, capacities and infrastructure that exist between countries, and in many cases within countries and that will enable opportunities for effective ICT uses.

Agreement on a substantively broad and procedurally inclusive approach to Internet governance, the reform of existing governance mechanisms in accordance with the Geneva principles, and the creation of a new forum to promote multi-stakeholder dialogue, analysis, trend monitoring, and capacity building in the field of Internet governance.

Ensuring that our human-centred vision of the 'Information Society', framed by a global commitment to human rights, social justice and inclusive and sustainable development, is present throughout the implementation phase.

Achieving a change of tide in perceptions and practices of participatory decision-making. We saw the WSIS as a milestone from which the voluntary and transparent participation of Civil Society would become more comprehensive and integrated at local, national, regional and global levels of governance and decision making.

Agreement on strong commitment to the centrality of human rights, especially the right to access and impart information and to individual privacy.

Civil Society affirms that, facing very limited resources, it has contributed positively to the WSIS process, a contribution that could have been even greater had the opportunity been made available for an even more comprehensive participation on our part. Our contribution will continue beyond the summit. It is a contribution that is made both through constructive engagement and through challenge and critique.

While we value the process and the outcomes, we are convinced much more could have been achieved. We have taken a month after the closure of the Tunis Summit to discuss the outcomes and the process of WSIS. We built on our Geneva 2003 Civil Society Summit Declaration "Shaping Information Societies for Human Needs," and we evaluated the experiences and lessons learned in the four years of WSIS I and WSIS II. This statement was developed in a global online consultation process. It is presented as Civil Society's official contribution to the summit outcomes.

The issues of greatest concern to Civil Society are addressed in sections II and III of this statement. For most of these items, minor achievements in the outcomes from WSIS were offset by major shortcomings, with much remaining to be done. Some of our greatest concerns involve what we consider to be insufficient attention or inadequate recommendations concerning people-centred issues such as the degree of attention paid to human rights and freedom of expression, the financial mechanisms for the promotion of development that was the original impetus for the WSIS process, and support for capacity building. In section IV, we lay out the first building blocks of Civil Society's "Tunis Commitment." Civil Society has every intention to remain involved in the follow-up and implementation processes after the Tunis summit. We trust governments realize that our participation is vital to achieve a more inclusive and just Information Society.

II. Issues Addressed During the Tunis Phase of WSIS

Social Justice, Financing and People-Centred Development

The broad mandate for WSIS was to address the long-standing issues in economic and social development from the newly emerging perspectives of the opportunities and risks posed by the revolution in Information and Communications Technologies (ICTs). The summit was expected to identify and articulate new development possibilities and paradigms being made possible in the Information Society, and to evolve public policy options for enabling and realising these opportunities. Overall, it is impossible not to conclude that WSIS has failed to live up to these expectations. The Tunis phase in particular, which was presented as the "summit of solutions," did not provide concrete achievements to meaningfully address development priorities.

While the summit did discuss the importance of new financing mechanisms for ICT for Development (ICTD), it failed to recognize that ICTD presents a challenge beyond that of traditional development financing. Nor did the Tunis fully comprehend that new means and sources of financing and the exploration of new models and mechanisms are required.

Investments in ICTD—in infrastructure, capacity building, appropriate software and hardware and in developing applications and services—underpin all other processes of development innovation, learning and sharing, and should be seen in this light. Though development resources are admittedly scarce and have to be allocated with care and discretion, ICTD financing should not be viewed as directly in competition with the financing of other developmental sectors. Financing ICTD should be considered a priority at both national and international levels, with specific approaches to each country according to its level of development and with a long-term perspective adapted to a global vision of development and sharing within the global community.

Financing ICTD requires social and institutional innovation, with adequate mechanisms for transparency, evaluation, and follow-up. Financial resources need to be mobilised at all levels—local, national and international, including through the realization of ODA commitments agreed to in the Monterrey Consensus and including assistance to programs and activities whose short-term sustainability cannot be immediately demonstrated because of the low level of resources available as their starting point.

Internet access, for everybody and everywhere, especially among disadvantaged populations and in rural areas, must be considered as a global public good. In many cases market approaches are unlikely to address the connectivity needs of particularly disadvantaged regions and populations. In many such areas, initial priority may need to be given to the provision of more traditional ICTs—radio, TV, video and telephony—while the conditions are developed for ensuring the availability of complete Internet connectivity. Infostructure and development often require attention to the development of more traditional infrastructure as well such as roads and electricity.

While the summit in general has failed to agree on adequate funding for ICTD, Civil Society was able to introduce significant sections in the Tunis Commitment (paragraph 35) and in the Tunis Agenda (paragraph 21) on the importance of public policy in mobilizing resources for financing. This can serve as a balance to the market-based orientation of much of the text on financing.

The potential of ICT as tools for development, and not merely tools for communication, by now should have been realised by all states. National ICT strategies should be closely related to national strategies for development and poverty eradication. Aid strategies in developed countries should include clear guidelines for the incorporation of ICT into all aspects of development. In this way ICTs should be integrated into general development assistance and in this way contribute to the mobilisation of additional resources and an increase in the efficiency of development assistance.

We welcome the launch of the Digital Solidarity Fund (DSF) in March 2005 and take note of the support it got both from the United Nations and the Tunis Summit. Nevertheless, taking into account that the DSF was established on a voluntary basis, we are concerned that there are no clear commitments from governments and the private sector to provide the needed material support to ensure the success of this fund. We invite all partners from the governmental and the private sector to commit themselves to the so-called "Geneva Principle" where each ICT contract concluded by a public administration with a private company includes a one percent contribution to the DSF. We particularly encourage local and regional administrations to adopt this principle and welcome the relevant statement made by the World Summit of Cities and Local Authorities in Bilbao, November 2005, on the eve of WSIS II.

Human Rights

The Information Society must be based on human rights as laid out in the Universal Declaration of Human Rights. This includes civil and political rights, as well as social, economic and cultural rights. Human rights and development are closely linked. There can be no development without human rights, no human rights without development.

This has been affirmed time and again, and was strongly stated in the Vienna World Conference on Human Rights in 1993. It was also affirmed in the WSIS 2003 Declaration of Principles. All legislation, policies, and actions involved in developing the global Information Society must respect, protect and promote human rights standards and the rule of law.

Despite the Geneva commitment to an Information Society respectful of human rights, there is still a long way to go. A number of human rights were barely addressed in the Geneva Declaration of Principles. This includes the cross-cutting principles of non-discrimination, gender equality, and workers' rights. The right to privacy, which is the basis of autonomous personal development and thus at the root of the exertion of many other fundamental human rights, is only mentioned in the Geneva Declaration as part of "a global culture of cyber-security." In the Tunis Commitment, it has disappeared, to make room for extensive underlining of security needs, as if privacy were a threat to security, whereas the opposite is true: privacy is an essential requirement for security. The summit has also ignored our demand that the principle of the privacy and integrity of the vote be ensured if and when electronic voting technologies are used.

Other rights were more explicitly addressed, but are de facto violated on a daily basis. This goes for freedom of expression, freedom of information, freedom of association and assembly, the right to a fair trial, the right to education, and the right to a standard of living adequate for the health and well-being of the individual and his or her family.

Furthermore, as the second WSIS phase has amplified, a formal commitment is one thing, implementation is something else. Side events open to the general public were organised by civil society both at the Geneva and Tunis Summit, consistent with a long tradition in the context of UN summits. In Tunis, the initiative by parts of civil society to organize a "Citizens' Summit on the Information Society" was prevented from happening. At the Geneva Summit, the "We Seize" event was closed down and then reopened. This is a clear reminder that though governments have signed on to human rights commitments, fundamental human rights such as freedom of expression and freedom of assembly cannot be taken for granted in any part of the world.

The summit has failed to define mechanisms and actions that would actively promote and protect human rights in the Information Society. Post-WSIS there is an urgent need to strengthen the means of human rights enforcement, to ensure the embedding of human rights proofing in national legislation and practises, to strengthen education and awareness raising in the area of rights-based development, to transform human rights standards into ICT policy recommendations, and to mainstream ICT issues into the global and regional human rights monitoring system—in summary: To move from declarations and commitments into action. Toward this end, an independent commission should be established to review national and international ICT regulations and practices and their compliance with international human rights standards. This commission should also address the potential applications of ICTs for the realization of human rights in the Information Society.

Internet Governance

Civil Society is pleased with the decision to create an Internet Governance Forum (IGF), which it has advocated for since 2003. We also are pleased that the IGF will have sufficient scope to deal with the issues we believe must be addressed, most notably the conformity of existing arrangements with the Geneva Principles, and other cross-cutting or multidimensional issues that cannot be optimally dealt with within current arrangements. However, we reiterate our concerns that the Forum must not be anchored in any existing specialized international organization, meaning that its legal form, finances, and professional staff should be independent. In addition, we reiterate our view that the forum should be more than a place for dialogue. As was recommended by the WGIG Report, it should also provide expert analysis, trend monitoring, and capacity building, including in close collaboration with external partners in the research community.

We are concerned about the absence of details on how this forum will be created and on how it will be funded. We insist that the modalities of the IGF be determined in full cooperation with Civil Society. We emphasize that success in the forum, as in most areas of Internet governance, will be impossible without the full participation of Civil Society. By full participation we mean much more than playing a mere advisory role. Civil Society must be able to participate fully and equally both in plenary and in any working or drafting group discussions, and must have the same opportunities as other stakeholders to influence agendas and outcomes.

The Tunis Agenda addressed the issue of political oversight of critical Internet resources in its paragraphs 69 to 71. This, in itself, is an achievement. It is also important that governments recognized the need for the development of a set of Internet-related public policy principles that would frame political oversight of Internet resources. These principles must respect, protect and promote human rights as laid down in international human rights treaties, ensure equitable access to information and online opportunities for all, and promote development.

It is important that governments have established that developing these principles should be a shared responsibility. However, it is very unfortunate that the Tunis Agenda suggests that governments are only willing to share this role and responsibility among themselves, in cooperation with international organisations. Civil Society remains strongly of the view that the formulation of appropriate and legitimate public policies pertaining to Internet governance requires the full and meaningful involvement of non-governmental stakeholders.

With regard to paragraph 40 of the Tunis Agenda, we are disappointed that there is no mention that efforts to combat cyber-crime need to be exercised in the context of checks and balances provided by fundamental human rights, particularly freedom of expression and privacy.

With regard to paragraph 63, we believe that a country code Top Level Domain (ccTLD) is a public good both for people of the concerned country or economy and for global citizens who have various linkages to particular countries. While we recognize the important role of governments in protecting the ccTLDs that refer to their countries or economies, this role must be executed in a manner that respects human rights as expressed in existing international treaties through a democratic, transparent and inclusive process with full involvement of all stakeholders.

To ensure that development of the Internet and its governance takes place in the public interest, it is important for all stakeholders to better understand how core Internet governance functions—as, for example, DNS management, IP address allocation, and others—are carried out. It is equally important that these same actors understand the linkages between broader Internet governance and Internet related matters such as cyber-crime, Intellectual Property Rights, e-commerce, e-government, human rights and capacity building and economic development. The responsibility of creating such awareness should be shared by everyone, including those at present involved in the governance and development of the Internet and emerging information and communication platforms. Equally it is essential that as this awareness develops in newer users of the Internet, older users must be open to the new perspectives that will emerge.

Global Governance

A world that is increasingly more connected faces a considerable and growing number of common issues which need to be addressed by global governance institutions and processes. While Civil Society recognises that there are flaws and inefficiencies in the United Nations system that require urgent reform, we believe strongly that it remains a most legitimate inter-governmental forum, where rich and poor countries have the same rights to speak, participate, and make decisions together.

We are concerned that during the WSIS it emerged that some governments, especially from developed countries, lack faith in, and appear to be unwilling to invest authority and resources in the present multilateral system, along with concerted efforts to further improve it. We also regret that debates on creating private-public partnerships and new para-institutions within the United Nations have over-shadowed the overall discussion on bridging the digital divide, which in turn has to be linked to a deep reform of the UN and the global economic system.

In our understanding, summits take place precisely to develop the principles that will underpin global public policy and governance structures; to address critical issues, and to decide on appropriate responses to these issues. Shrinking global public policy spaces raise serious questions concerning the kind of global governance toward which we are heading, and what this might mean for people who are socially, economically and politically marginalised: precisely those people who most rely on public policy to protect their interests.

Participation

In the course of four years, as a result of constant pressure from Civil Society, improvements in Civil Society participation in these processes have been achieved, including speaking rights in official plenaries and sub-committees, and ultimately rights to observe in drafting groups. The UN Working Group on Internet Governance created an innovative format where governmental and Civil Society actors worked on an equal footing and Civil Society actually carried a large part of the drafting load.

Due to the pressure of time and the need of governments to interact with Civil Society actors in the Internet Governance field, the resumed session of

PrepComIII was in fact the most open of all. We would like to underline that this openness, against all odds, contributed to reaching consensus.

WSIS has demonstrated beyond any doubt the benefits of interaction between all stakeholders. The innovative rules and practices of participation established in this process will be fully documented to provide a reference point and a benchmark for participants in UN organizations and processes in the future.

Civil Society thanks those governments and international bodies that greatly supported our participation in the WSIS process. We hope and expect that these achievements are taken further and strengthened, especially in more politically contested spaces of global policymaking such as those concerning intellectual property rights, trade, environment, and peace and disarmament.

We note that some governments from developing countries were not actively supportive of greater observer participation believing that it can lead to undue dominance of debate and opinions by international and developed countries' Civil Society organisations and the private sector. We believe that to change this perception, efforts should be engaged in to strengthen the presence, independence and participation of Civil Society constituencies in and from their own countries.

As for the period beyond the summit, the Tunis documents clearly establish that the soon-to-be created Internet Governance Forum, and the future mechanisms for implementation and follow-up (including the revision of the mandate of the ECOSOC Commission on Science and Technology for Development) must take into account the multi-stakeholder approach.

We want to express concern at the vagueness of text referring to the role of Civil Society. In almost every paragraph talking about multi-stakeholder participation, the phrase "in their respective roles and responsibilities" is used to limit the degree of multi-stakeholder participation. This limitation is due to the refusal of governments to recognize the full range of the roles and responsibilities of Civil Society. Instead of the reduced capabilities assigned in paragraph 35C of the Tunis Agenda that attempt to restrict Civil Society to a community role, governments should have at minima referred to the list of Civil Society roles and responsibilities listed in the WGIG report. These are:

Awareness raising and capacity building (knowledge, training, skills sharing);

Promote various public interest objectives;

Facilitate network building;

Mobilize citizens in democratic processes;

- Bring perspectives of marginalized groups including, for example, excluded communities and grassroots activists;
- Engage in policy processes;
- Bring expertise, skills, experience and knowledge in a range of ICT policy areas contributing to policy processes and policies that are more bottom-up, people-centred and inclusive;
- Research and development of technologies and standards;
- Development and dissemination of best practices;
- Helping to ensure that political and market forces are accountable to the needs of all members of society;
- Encourage social responsibility and good governance practice;
- Advocate for development of social projects and activities that are critical but may not be 'fashionable' or profitable;
- Contribute to shaping visions of human-centred information societies based on human rights, sustainable development, social justice and empowerment.

Civil Society has reason for concern that the limited concessions obtained in the last few days before the summit, from countries that previously refused the emergence of a truly multi-stakeholder format, will be at risk in the coming months. Civil Society actors therefore intend to remain actively mobilized. They need to proactively ensure that not only the needed future structures be established in a truly multi-stakeholder format, but also that the discussions preparing their mandates are conducted in an open, transparent and inclusive manner, allowing participation of all stakeholders on an equal footing. Civil Society hopes to be given the means to ensure all its representatives from different regions, languages and cultures, from developed and developing countries, can fully participate.

III. Issues Addressed in the Geneva and Tunis Phases

Gender Equality

Equal and active participation of women is essential, especially in decision-making. This includes all forums that will be established in relation to WSIS and the issues it has taken up. With that, there is a need for capacity building that is focussed on women's engagement with the shaping of an Information Society at all levels, including policy making on infrastructure development, financing, and technology choice.

There is a need for real effort and commitment to transforming the masculinist culture embedded within existing structures and discourses of the Information Society which serves to reinforce gender disparity and inequality. Without full, material and engaged commitment to the principle of gender equality, women's empowerment and non-discrimination, the vision of a just and equitable Information Society cannot be achieved.

Considering the affirmation of unequivocal support for gender equality and women's empowerment expressed in the Geneva Declaration of Principles and paying careful attention to Paragraph 23 of the Tunis Commitment, all government signatories must ensure that national policies, programs and strategies developed and implemented to build a people-centred, inclusive and development-oriented Information Society demonstrate significant commitment to the principles of gender equality and women's empowerment.

We emphasise that financial structures and mechanisms need to be geared towards addressing the gender divide, including the provision of adequate budgetary allocations. Comprehensive gender-disaggregated data and indicators have to be developed at national levels to enable and monitor this process. We urge all governments to take positive action to ensure that institutions and practices, including those of the private sector, do not result in discrimination against women. Governments that are parties to the UN Convention on the Elimination of All Forms of Discrimination against Women (CEDAW) are in fact bound to this course of action.

Culture, Knowledge, and the Public Domain

Each generation of humankind is depending upon its predecessors to leave them with a liveable, sustainable and stable environment. The environment we were discussing throughout the WSIS is the public domain of global knowledge. Like our planet with its natural resources, that domain is the heritage of all humankind and the reservoir from which new knowledge is created. Limited monopolies, such as copyrights and patents were originally conceived as tools to serve that public domain of global knowledge to the benefit of humankind. Whenever society grants monopolies, a delicate balance must be struck: Careless monopolization will make our heritage unavailable to most people, to the detriment of all.

It has become quite clear that this balance has been upset by the interests of the rights-holding industry as well as the digitalization of knowledge. Humankind now has the power to instantaneously share knowledge in real-time, without loss, and at almost no cost. Civil Society has worked hard to defend that ability for all of humankind.

Free software is an integral part of this ability: Software is the cultural technique and most important regulator of the digital age. Access to it determines who may participate in a digital world. While in the Geneva phase, WSIS has recognised the importance of free software, it has not acted upon that declaration and this recognition faded in the Tunis phase. In the Tunis Commitment, free software is presented as a software model next to proprietary software, but paragraph 29 reiterates "the importance of proprietary software in the markets of the countries." This ignores that a proprietary software market is always striving towards dependency and monopolization, both of which are detrimental to economy and development as a whole. Proprietary software is under exclusive control of and to the benefit of its proprietor. Furthermore: Proprietary software is often written in modern sweat-shops for the benefit of developed economies, which are subsidized at the expense of developing and least-developed countries in this way.

While WSIS has somewhat recognised the importance of free and open source software, it has not asserted the significance of this choice for development. It is silent on other issues like open content (which goes beyond open access in the area of academic publications), new open telecom paradigms and community-owned infrastructure as important development enablers.

The WSIS process has failed to introduce cultural and linguistic diversity as a cross-cutting issue in the Information Society. The Information Society and its core elements—knowledge, information, communication and the information and communication technologies (ICT) together with related rules and standards—are cultural concepts and expressions. Accordingly, culturally defined approaches, protocols, proceedings and obligations have to be respected and culturally appropriate applications developed and promoted. In order to foster and promote cultural diversity it must be ensured that no one has to be a mere recipient of Western knowledge and treatment. Therefore development of the cultural elements of the Information Society must involve strong participation by all cultural communities. The WSIS has failed to recognize the need for developing knowledge resources to shift the current lack of diversity, to move from the dominant paradigm of over-developed nations and cultures to the need for being open to learning and seeing differently.

Indigenous peoples, further to self-determination and pursuant to their traditional and customary laws, protocols, rules and regulations, oral and written, provide for the access, use, application and dissemination of traditional and cultural knowledge, oral histories, folklore and related customs and practices. WSIS has failed to protect these from exploitation, misuse and appropriation by third parties. As a result, the traditional knowledge, oral histories, folklore and related customs, practices and representations have been and continue to be exploited by both informal and formal (being copyright, trademark and patent) means, with no benefits to the rightful Indigenous holders of that knowledge.

Education, Research, and Practice

If we want future generations to understand the real basis of our digital age, freedom has to be preserved for the knowledge of humankind: free software, open courseware and free educational as well as scientific resources empower people to take their life into their own hands. If not, they will become only users and consumers of information technologies, instead of active participants and well informed citizens in the Information Society. Each generation has a choice to make: Schooling of the mind and creativity, or product schooling? Most unfortunately, the WSIS has shown a significant tendency towards the latter.

We are happy that universities, museums, archives, libraries have been recognized by WSIS as playing an important role as public institutions and with the community of researchers and academics. Unfortunately, telecenters are missing in the WSIS documents. Community informatics, social informatics, telecenters and human resources such as computer professionals, and the training of these, have to be promoted, so that ICT serves training and not training serves ICT. Thus special attention must be paid to supporting sustainable capacity building with a specific focus on research and skills development. In order to tackle development contexts training should have a sociological focus too and not be entirely technologically framed.

Problems of access, regulation, diversity and efficiency require attention to power relations both in the field of ICT policy-making and in the everyday uses of ICT. Academic research should play a pivotal role in evaluating whether ICT meets and serves the individuals' and the public's multiple needs and interests—as workers, women, migrants, racial, ethnic and sexual minorities, among others—across very uneven information societies throughout the world. Furthermore, because power relations and social orientations are often embedded in the very designs of ICT, researchers should be sensitive to the diverse and multiple needs of the public in the technological design of ICT. Similarly, educators at all levels should be empowered to develop curricula that provide or contribute to training for people not only as workers and consumers using ICT, but also in the basic science and engineering of ICT, in the participatory design of ICT by communities with computing professionals, the critical assessment of ICT, the institutional and social contexts of their development and implementation, as well as their creative uses for active citizenship. Young people—given their large numbers, particularly in developing countries, and enthusiasm and expertise in the use of ICTs—remain an untapped resource as initiators of peer-to-peer learning projects at the community and school levels. These issues have largely been ignored by WSIS.

The actors that need to be involved in the process of making this vision a reality are the professionals and researchers, the students and their families, the support services and human resources of the resources centres, politicians at all levels, social organizations and NGOs, but also the private sector. However, in the teaching profession, it is necessary to recognize and accept the need for learning and evolution with regards to ICT.

We emphasize the special role that the computing, information science, and engineering professions have in helping to shape the Information Society to meet human needs. Their education must encourage socially-responsible practices in the design, implementation, and operation of ICT. The larger Information Society has an equally important and corresponding role to play by participating in the design of ICT. We, therefore, encourage increased cooperation between the computing, information science, and engineering professions and end-users of ICTs, particularly communities.

We furthermore have repeatedly underlined the unique role of ICT in socio-economic development and in promoting the fulfilment of internationally agreed development goals, including those contained in the Millennium Declaration. This is not least true in the reference to access to information and universal primary education. To secure the fulfilment of these goals, it is of key importance that the issue of ICT as tools for the improvement of education is also incorporated in the broader development strategies at both national and international levels.

Media

We are pleased that the principle of freedom of expression has been reaffirmed in the WSIS II texts and that they echo much of the language of Article 19 of the Universal Declaration of Human Rights. While we note that the Tunis Commitment recognises the place of the media in a new Information Society, this should never have been in question.

In the future, representatives of the media should be assured a place in all public forums considering development of the Internet and all other relevant aspects of the Information Society. As key actors in the Information Society, the media must have a place at the table, and this must be fully recognized both by governments and by Civil Society itself.

While recognizing media and freedom of expression, the WSIS documents are weak on offering support for developing diversity in the media sector and for avoiding a growing concentration and uniformity of content. They specifically neglect a range of projects and initiatives which are of particular value for Civil Society and which need a favourable environment: Community media, telecenters, grassroots and Civil Society-based media. These media empower people for independent and creative participation in knowledge-building and information-sharing. They represent the prime means for large parts of the world population to participate in the Information Society and should be an integral part of the public policy implementation of the goals of the Geneva Declaration, which refers to the promotion of the diversity of media and media ownership.

The WSIS documents also mostly focus on market-based solutions and commercial use. Yet the Internet, satellite, cable and broadcast systems all utilize public resources, such as airwaves and orbital paths. These should be managed in the public interest as publicly owned assets through transparent and accountable regulatory frameworks to enable the equitable allocation of resources and infrastructure among a plurality of media including community media. We reaffirm our commitment that commercial use of these resources begins with a public interest obligation.

Universal Design and Assistive Technologies

We are pleased to note that WSIS has identified the fact that ICT Design is the core issue of the digital divide for persons with disabilities. The Tunis Agenda for the Information Society clearly states in its paragraph 90e "paying special attention to the formulation of universal design concepts and the use of assistive technologies that promote access for all persons, including those with disabilities." Due to great efforts of all stakeholders, in particular of those with disabilities, we recognize significant advancement in the common understanding on the digital divide of persons with disabilities and strategies to achieve the targets set out in the Geneva Plan of Action to be achieved by ICT development with the Universal Design Concept in combination with Assistive Technologies that meet specific requirements of persons with disabilities.

In terms of equal opportunities for the participation of persons with disabilities in WSIS, the process of that was addressed in Geneva Declaration of the Global Forum on Disability in the Information Society in Geneva, we are grateful for all efforts extended by the summit organizers, who established a focal point for participants with disabilities at the last stage. However, there is still a lot to do to ensure equal participation of persons with disabilities in the WSIS Action Plan implementation process.

We call upon all governments, private sectors, civil society and international organizations to make the implementation, evaluation and monitoring of all WSIS documents, both from the first and second phase, inclusive to persons with disabilities. We urge that persons with disabilities be included in all aspects of designing, developing, distributing and deploying of appropriate strategies for ICT, including information and communication services, so as to ensure accessibility for persons with disabilities, taking into account the universal design principle and the use of assistive technologies. We request that any international, regional and national development program, funding or assistance aimed to achieve the inclusive information society be made disability-inclusive, both through mainstreaming and disability-specific approaches. We urge all governments to support the process of negotiation, adoption, ratification and implementation of the International Convention on the Rights of Persons with Disabilities, in particular through enactment of national legislation, as it contains strong elements concerning information and communication accessibility for persons with disabilities.

Health Information

Access to health information and knowledge is essential to collective and individual human development and has been identified as a critical factor in the public physical and mental health care crises around the world. The WSIS process has neglected to recognize that health is a cross-cutting issue and that health systems must include a holistic approach which is integral to the promotion of physical and mental health and the prevention and treatment of physical and mental illness for all people and to achieve the Millennium Development Goals (MDGs).

It is important to recognize that health expertise and scientific knowledge is essential to aid disease stricken, as well as traumatized populations affected by war, terrorism, disaster and other events, and further that the implementation of ICT systems for physical and mental health information and services must be a two-way path recognizing cultural and community norms and values.

It is essential that health care specialists, practitioners, and consumers participate in the development of public policy addressing privacy and related issues regarding physical and mental health information affecting information and delivery systems.

Children and Young People in the Information Society

In WSIS Phase I, the Geneva Declaration of Principles explicitly acknowledged young people, in paragraph 11, as the "future workforce and leading creators and earliest adopters of ICTs" and that to fully realize this end, youth must be "empowered as learners, developers, contributors, entrepreneurs and decision-makers." The Tunis Commitment in paragraph 25 reaffirmed the strategic role of youth as stakeholders and partners in creating an inclusive Information Society. This recognition is further supported by paragraph 90 of the Tunis Agenda. However, we are concerned as to how key decision-makers from governments, the business community and Civil Society will realize this commitment when the existing structures are not open for genuine, full and effective participation by youth. None of the Tunis documents, specifically in the post-WSIS implementation and follow-up parts, clearly defines how youth shall be "actively engaged in innovative ICT-based development programmes and...in e-strategy processes," as paragraph 25 states. In this regard, we call upon governments, both national and local, and the proponents of the Digital Solidarity Fund, to engage young people as digital opportunities are created and national e-strategies developed. Youth must be tapped as community leaders and volunteers for ICT for Development projects and be consulted in global and national ICT policy-making processes and formulation.

While we support the great opportunities that ICTs offer children and young people, paragraph 90q of the Tunis Agenda and paragraph 24 of the Tunis Commitment outline the potential dangers that children and young people face in relation to ICTs. For this reason, paragraph 92 of the Tunis Agenda encourages all governments to support an easy to remember, free of charge, national number for all children in need of care and protection. However, we had hoped that WSIS would have encouraged every stakeholder to support a more comprehensive proposal that ensured that every child, especially those that are marginalized and disadvantaged, has free access to ICTs, including but not limited to, toll free landlines, mobile telephones and Internet connection. In this regard, strategies should be developed that allow children and young people to reap the benefits that ICTs offer by making ICT an integral part of the formal and informal education sectors. There should also be strategies that protect children and young people from the potential risks posed by new technologies, including access to inappropriate content, unwanted contact and commercial pressures, particularly with regards to pornography, pedophilia and sexual trafficking, while fully respecting human rights standards on freedom of expression. We are committed to work in the WSIS follow-up process towards a world where telecommunication allows children and young people to be heard one-by-one and, through their voices, to fulfil their rights and true potential to shape the world.

Ethical Dimensions

The Tunis texts would have clearly been stronger if the aspects of the Information Society being people-centred, human rights-based and sustainable development-oriented were seen as the ethical point of departure in human relationships and community building and equally in bodies of international agreements. These ethical dimensions are foundational to a just, equitable and sustainable information and knowledge society.

Geneva identified the ethical values of respect for peace and the fundamental values of freedom, equality, solidarity, tolerance, shared responsibility, and respect for nature as enunciated in the Millennium Declaration. Tunis should have improved on these by including the principles of trust, stewardship and shared responsibility together with digital solidarity. The technologies we develop, and the solidarities we forge, must build relationships and strengthen social cohesion

Human rights conventions, for example, are critically important in evaluating ICTs so that they are tools to enable just and peaceable conditions for humanity. But Tunis failed to point in this direction. It did not, for example, restate what Geneva considered as acts inimical to the Information Society such as racism, intolerance, hatred, violence and others.

The strong emphasis on technology in the Tunis texts must not eclipse the human being as the subject of communication and development. Our humanity rests in our capacity to communicate with each other and to create community. It is in the respectful dialogue and sharing of values among peoples, in the plurality of their cultures and civilizations, that meaningful and accountable communication thrives. The Tunis texts did not give clear indications on how this can happen.

In an age of economic globalization and commodification of knowledge, the ethics and values of justice, equity, participation and sustainability are imperative. Beyond Tunis, all stakeholders must be encouraged to weave ethics and values language into the working on semantic web knowledge structures. Communication rights and justice are about making human communities as technology's home and human relationships as technology's heart.

IV. Where to Go from Here—Our Tunis Commitment

Civil Society is committed to continuing its involvement in the future mechanisms for policy debate, implementation and follow-up on Information Society issues. To do this, Civil Society will build on the processes and structures that were developed during the WSIS process.

Element One: Evolution of Our Internal Organization

Civil Society will work on the continued evolution of its current structures. This will include the use of existing thematic caucuses and working groups, the possible creation of new caucuses, and the use of the Civil Society Plenary, the Civil Society Bureau, and the Civil Society Content and Themes Group. We will organise, at a date to be determined, to launch the process of creating a Civil Society charter.

Element Two: Involvement in the Internet Governance Forum

The Civil Society Internet Governance Caucus will actively participate in and support the work of the Internet Governance Forum (IGF), and is exploring ways to enhance its working methods and its engagement with relevant stakeholders, especially the research community, to these ends. In addition, the caucus is considering the creation of a new Working Group that will make recommendations on the IGF, and other Civil Society caucuses, and individual Civil Society Working Groups will develop ideas for and participate in the IGF as well.

Element Three: Involvement in Follow-up and Implementation

In order to ensure that future implementation and follow-up mechanisms respect the spirit and letter of the Tunis documents and that governments uphold the commitments they have made during this second phase of the WSIS, Civil Society mechanisms will be used and created to ensure:

> the proactive monitoring of and participation in the implementation of the Geneva Plan of Action and the Tunis Agenda at the national level;

> a structured interaction with all UN agencies and international organisations and regional as well as national mechanisms for follow-up, to ensure that they integrate the WSIS objectives in their own work

plans, and that they put in place effective mechanisms for multi-stakeholder interaction, as mentioned in paragraphs 100 and 101 of the Tunis Agenda;

that the Information Society as a complex social political phenomenon is not reduced to a technology-centred perspective. The ECOSOC Commission on Science and Technology for Development will have to change significantly its mandate and composition to adequately address the need for being an effective follow-up mechanism for WSIS while re-affirming its original mission of developing science and technology, in addition to ICT, for the development objectives of poor countries;

not only that the reformed Commission on Science and Technology for Development truly becomes a multi-stakeholder commission for the Information Society, but also that the process to revise its mandate, composition and agenda is done in a fully open and inclusive manner.

Element Four: Lessons Learned for the UN System in General

We see the WSIS process as an experience to be learned from for the overall UN system and related processes. We will therefore work with the United Nations and all stakeholders on:

developing clearer and less bureaucratic rules of recognition for accrediting Civil Society organisations in the UN system, for instance in obtaining ECOSOC status and summit accreditation, and to ensure that national governmental recognition of Civil Society entities is not the basis for official recognition in the UN system; and

ensuring that all future summit processes be multi-stakeholder in their approach, allowing for appropriate flexibility. This would be achieved either by recognition of precedents set in summit processes, or by formulating a rules of procedure manual to guide future summit processes and day-to-day Civil Society interaction with the international community.

Element Five: Outreach to Other Constituencies

The civil society actors who actively participated in the WSIS process are conscious that the Information Society, as its name suggests, is a society-wide phenomenon, and that advocacy on Information Society issues need to include every responsible interest and group. We therefore commit ourselves in the post-WSIS period to work to broaden our reach to include different Civil Society constituencies that for various reasons have not been active in the WSIS process; may have shown scepticism over the role of ICT in their core areas of activity; or for other reasons have remained disengaged from the Information Society discourse.

Bibliography

Key Documents Produced by Civil Society Within the Framework of the WSIS

WSIS Civil Society Plenary. *Much More Could Have Been Achieved: Civil Society Statement on the World Summit on the Information Society*, December 18, 2005 Revision 1–December, 23 2005.
http://www.worldSummit2003.de/download_en/WSIS-CS-Summit-statement-rev1-23-12-2005-en.pdf

WSIS Civil Society Plenary. *Shaping Information Societies for Human Needs: Civil Society Declaration to the World Summit on the Information Society*, December 8, 2003.
http://www.worldsummit 2005.de/download_en/WSIS-CS-Dec-25-Feb-04-en.pdf

Declarations of WSIS (Phase I & Phase II)

Geneva Declaration of Principles (WSIS-03/GENEVA/DOC/4-E), December 12, 2003.
  http://www.itu.int/wsis/docs/geneva/official/dop.html
Geneva Plan of Action (WSIS-03/GENEVA/DOC/5-E), December 12, 2003.
  http://www.itu.int/wsis/docs/geneva/official/poa.html
Tunis Agenda for the Information Society (WSIS-05/TUNIS/DOC/6 (rev. 1), November 18, 2005.
  http://www.itu.int/wsis/docs2/tunis/off/6rev1.doc
Tunis Commitment (WSIS-05/TUNIS/DOC/7), November 18, 2005.
  http://www.itu.int/wsis/docs2/tunis/off/7.doc

General Analyses of the WSIS

Accuosto, Pablo and Niki Johnson. *Financing the Information Society in the South: A Global Public Goods Perspective*. Working Paper, Instituto del Tercer Mundo: Montevideo, 2004. http://rights.apc.org/documents/financing.pdf

Afonso, Carlos A. *Internet Governance: a Review in the Context of the WSIS Process*. Prepared for Instituto del Tercer Mundo (ITeM): Montevideo, July 2005. http://wsispapers.choike.org/Internet_governance.pdf

Association for Progressive Communications. *Pushing and Prodding, Goading and Hand-holding. Reflection from the Association for Progressive Communications (APC) at the Conclusion of the World Summit on the Information Society*. Montevideo, February 14, 2006. http://www.apc.org/en/system/files/apc_wsis_reflection_0206.pdf

Bendrath, Ralf and Rik Panganiban. *How Was the Summit? A Helpful List in Case Your Friends (Or Any Reporters) Ask You*, December 16, 2003, http://www.worldSummit2003.de/en/web/577.htm

Bertola, Vittorio. "Oversight and Multiple Root Servers." In William Drake (ed.), *Reforming Internet Governance: Perspectives from the Working Group on Internet Governance (WGIG)*. New York: UN ICT Task Force, 2005.

Burch, Sally. "Global Media Governance: Reflections from the WSIS Experience." *Media Development* 2004 (1).

Busaniche, Beatriz. "Civil Society in the Carousel: Who Wins, Who Loses and Who Is Forgotten by the Multi-stakeholder Approach?" In Olga Drossou and Heike Jensen (eds.), *Visions in Process II*. Berlin: The Heinrich Böll Foundation, 2005. http://www.worldSummit2005.de/download_en/Visions-in-ProcessII(1).pdf

Cammaerts, Bart and Nico Carpentier, "The Unbearable Lightness of Full Participation in a Global Context: WSIS and Civil Society Participation." In Jan Servaes and Nico Carpentier (eds.), *Towards a Sustainable Information Society: Beyond WSIS*. Bristol: Intellect, 2005: 17–49.

Clerc, Alain. "Innovative Financial Mechanisms, Digital Solidarity and the 'Geneva Principle'." In Daniel Stauffacher and Wolfgang Kleinwächter (eds.), *The World Summit on the Information Society: Moving from the Past into the Future*. New York: The United Nations Information and Communication Technologies Task Force, 2005: 178–182.

Currie, Willie. *Creating Spaces for Civil Society in the WSIS: A Reply to Michael Gurstein*, December 22, 2005. http://www.worldSummit2003.de/en/web/848.htm

Dany, Charlotte. "The Impact of Participation: How Civil Society Organisations Contribute to the Democratic Quality of the UN World Summit on the Information Society." *TranState Working Paper* 43. Bremen: Collaborative Research Center, 2006. http://www.state.uni-bremen.de/pages/pubApBeschreibung.php?SPRACHE=en&ID=50

Drake, William. "Reforming Internet Governance: Fifteen Baseline Propositions." In Don MacLean (ed.). *Internet Governance: A Grand Collaboration*. New York: UN ICT Task Force, 2004: 122–161.

———. "Collective Learning in the World Summit on the Information Society. In Daniel Stauffacher and Wolfgang Kleinwächter (eds.), *The World Summit on the Information Society: Moving from the Past into the Future*. New York: UN ICT Task Force, 2005: 135–136.

———. (Ed.). *Reforming Internet Governance: Perspectives from the Working Group on Internet Governance (WGIG)*. New York: UN ICT Task Force, 2005.

Drossou, Olga and Heike Jensen. *Visions in Process II*. Germany: Heinrich Böll Foundation, 2005. http://www.worldSummit2003.de/download_en/Visions-in-ProcessII(1).pdf

Dutton, William H. 2005. *A New Framework for Taking Forward the Global Internet Governance Debate*. Draft Position Papers for the Oxford Internet Institute Discussion Forum, May 6th, 2005.

Frau-Meigs, Divina. "Civil Society's Involvement in the WSIS Process: Drafting the Alter-Agenda." In Jan Servaes and Nico Carpentier (eds.), *Towards a Sustainable Information Society: Beyond WSIS*. Bristol: Intellect, 2005: 81–96.

Gurstein, Michael. *Networking the Networked*, 2005. http://www.worldSummit2003.de/en/web/847.htm

Hamelink, Cees J. "Did WSIS Achieve Anything at All?" *Gazette: The International Journal for Communication Studies* 66 (3–4), 2004: 281–290.

Hintz, Arne. *Civil Society Media and Global Governance. Intervening into the World Summit on the Information Society*. Münster: Lit, 2009.

Hintz, Arne and Stefania Milan. "Towards a New Vision for Communication Governance? Civil Society Media at the World Social Forum and the World Summit on the Information Society." *Communication for Development and Social Change* 1 (1), 2007.

Klein, Hans. *ICANN Reform: Establishing the Rule of Law. A Policy Analysis Prepared for the WSIS.* Georgia Institute of Technology Internet and Public Policy Project, 2005.

———. "Understanding WSIS: An Institutional Analysis of the Summit." *Information Technologies and International Development* 1 (3-4), 2005: 3-13.

Kleinwächter, Wolfgang. "Beyond ICANN vs. ITU: Will WSIS Open New Territory for Internet Governance?" In Don MacLean (ed.), *Internet Governance: A Grand Collaboration*. New York: UN ICT Task Force, 2004: 31-52.

———. "WSIS: A New Diplomacy? Multistakeholder Approach and Bottom Up Policy Global ICT Governance." *Information Technology and International Development* 1 (3-4), 2004.

Kooiman, Jan. *Governing as Governance*. London; Thousand Oaks, Calif.: Sage, 2003.

Kummer, Markus. "Agree to Disagree: The Birth of the Working Group on Internet Governance." In Daniel Stauffacher and Wolfgang Kleinwächter (eds.), *The World Summit on the Information Society: Moving from the Past into the Future*. New York: UN ICT Task Force, 2005: 245-248.

———. 2007. "The Debate on Internet Governance: From Geneva to Tunis and Beyond." *Information Polity* 12 (1-2): 5-13.

Latzer, Michael. "Convergence Revisited: Toward a Modified Pattern of Communications Governance." *Convergence: The International Journal of Research into New Media Technologies* (15) 4, 2009: 411-426.

Lavoie, Manon and Peter Leuprecht (eds.). "Beyond WSIS: Incorporating Human Rights Perspectives into the Information Society Debate." *Revue Québécoise de Droit International* (special edition) 18 (1), 2005.

MacLean, Don. "A Brief History of the WGIG." In William Drake (ed.), *Reforming Internet Governance: Perspectives from the Working Group on Internet Governance (WGIG)*. New York: UN ICT Task Force, 2005: 9-24.

———. (Ed.). *Internet Governance: A Grand Collaboration*. New York: UN ICT Task Force, 2004.

Mansell, Robin and Kaarle Nordenstreng. "Great Media and Communication Debates: WSIS and the MacBride Report." *Information Technologies and International Development* 3 (4), 2008: 15-36.

McIver, William J. "Internet." In Marc Raboy and Jeremy Shtern (eds.), *Media Divides: Communication Rights and the Right to Communicate in Canada*. Vancouver: UBC Press, 2010.

McLaughlin, Lisa and Victor Pickard. "What Is Bottom Up about Global Internet Governance?" *Global Media and Communication* 1 (3), 2005: 357-373.

Moll, Marita and Leslie Regan Shade. "Vision Impossible? The World Summit on the Information Society." In Marita Moll and Leslie Regan Shade (eds.), *Seeking Convergence in Policy and Practice*. Ottawa: CCPA, 2004: 45-80.

Mueller, Milton, Brenden N. Kuerbris, and Christiane Pagé. "Democratizing Global Communication? Global Civil Society and the Campaign for Communication Rights in the Information Society." *International Journal of Communication* 1, 2007: 267-296.

Numminen, Asko. "Searching for Consensus." In Daniel Stauffacher and Wolfgang Kleinwächter (eds.), *The World Summit on the Information Society: Moving from the Past into the Future*. New York: UN ICT Task Force, 2005: 64-70.

Ó Siochrú, Sean. "Will the Real WSIS Please Stand-up? The Historic Encounter of the 'Information Society' and the 'Communication Society'." *Gazette: The International Journal for Communication Studies* 66 (3-4), 2004: 203-224.

———. "Civil Society Participation in the WSIS Process. Promises and Reality." *Continuum: Journal of Media & Cultural Studies* 18 (3), 2004: 330-344.

———. *Mapping Research Activities and Interactions against Global Communication Dynamics: The Case of the WSIS and Financing Mechanisms*, 2005. http://programs.ssrc.org/media/publications/OSiochru.6.FinalPaper.doc

Padovani, Claudia. "Debating Communication Imbalances from the MacBride Report to the World Summit on the Information Society: An Analysis of a Changing Discourse." *Global Media and Communication* 1 (3), 2005: 316-338.

———. "The World Summit on the Information Society: Setting the Communication Agenda for the 21st Century? An Ongoing Exercise." *Gazette: The International Journal for Communication Studies* 66 (3-4), 2004: 187-191.

Padovani, Claudia and Kaarle Nordenstreng. "From NWICO to WSIS: Another World Information and Communication Order?" *Global Media and Communication* 1 (3), 2005: 264-272.

Padovani, Claudia and Arjuna Tuzzi. "The WSIS as a World of Words: Building a Common Vision for the Communication Society?" *Continuum: Journal of Media & Cultural Studies* 18 (3), 2004: 330-344.

Peake, Adam. *Internet Governance and the World Summit on the Information Society (WSIS). A Report Prepared for the Association for Progressive Communications (APC)*. APC, 2004. http://www.apc.org/en/pubs/governance

Pickard, Victor. "Neoliberal Visions and Revisions in Global Communications Policy from NWICO to WSIS." *Journal of Communication Inquiry* 32 (2), 2007: 118-139.

Raboy, Marc. "The World Summit on the Information Society and its Legacy for Global Governance." *Gazette: The International Journal for Communication Studies* 66, 2004: 225-232.

———. "The WSIS as a Political Space in Global Media Governance." *Continuum: Journal of Media and Cultural Studies* 18 (3), 2004: 345-359.

———. "Globalization in the Information Society." In Daniel Stauffacher and Wolfgang Kleinwächter (eds.), *The World Summit on the Information Society: Moving from the Past into the Future*. New York: UN ICT Task Force, 2005: 131-134.

———. "Broadening Media Discourses: Global Media Policy—Defining the Field." *Global Media and Communication* 3 (3), 2007: 343-347.

Raboy, Marc and Normand Landry. *Civil Society, Communication and Global Governance: Issues from the World Summit on the Information Society*. New York: Peter Lang, 2005.

Servaes, Jan and Nico Carpentier (eds.). *Towards a Sustainable Information Society: Deconstructing WSIS*. Bristol: Intellect, 2006.

Stauffacher, Daniel and Wolfgang Kleinwächter (eds.). *The World Summit on the Information Society: Moving from the Past into the Future*. New York: UN ICT Task Force, 2005.

Thomas, Pradip. *CRIS and Global Media Governance: Communication Rights and Social Change*. Paper presented to the Social Change in the 21st Century Conference Centre for Social

Change Research, Queensland University of Technology, October 28, 2005. http://eprints.qut.edu.au/3480/1/3480.pdf

Working Group on Internet Governance (WGIG). *Report of the Working Group on Internet Governance*, June 2005. http://www.wgig.org/docs/WGIGREPORT.pdf

Zhao, Yuezhi. "Between a World Summit and a Chinese Movie: Visions of the 'Information Society'." *Gazette: The International Journal for Communication Studies* 66 (3-4), 2004: 275-280.

Related Topics on Communication and Global Governance

Anheier, Helmut, Marlies Glasius and Mary Kaldor. "Introducing Global Civil Society." In Helmut Anheier, Marlies Glasius and Mary Kaldor (eds.), *Global Civil Society Yearbook 2001*. London: Oxford University Press, 2001: 3-22.

Annan, Kofi A. *Report of the Secretary-General in Response to the Report of the Panel of Eminent Persons on United Nations-Civil Society Relations* (A/59/354). New York: United Nations General Assembly, September 13, 2004. http://daccess-dds-ny.un.org/doc/UNDOC/GEN/N04/507/26/PDF/N0450726.pdf?OpenElement

Ba, Alice D. and Matthew J. Hoffmann (eds.). *Contending Perspectives on Global Governance: Coherence, Contestation and World Order*. London; New York: Routledge, 2005.

Baird, Zoë. "Governing the Internet." *Foreign Affairs* 81 (6), 2002: 15-21.

Barnett, Michael and Raymond Duvall (eds.). *Power in Global Governance*. Cambridge, UK; New York: Cambridge University Press, 2005.

Barney, Darin. *Communication Technology*. Vancouver: UBC Press, 2005.

Baylis, J. and S. Smith. *The Globalization of World Politics. An Introduction to International Relations*. 3rd ed. Oxford: Oxford University Press, 2005.

Calabrese, Andrew. "The Promise of Civil Society: a Global Movement for Communication Rights." *Continuum: Journal of Media & Cultural Studies* 18 (3), 2004: 317-329.

Cameron, David and Janice Gross Stein. *Street Protests and Fantasy Parks: Globalization, Culture, and the State*. Vancouver: UBC Press, 2002.

Cardoso, Fernando Henrique, et al. *We the Peoples: Civil Society, the United Nations and Global Governance*. Report of the Panel of Eminent Persons on United Nations-Civil Society Relations (A/58/817). New York: United Nations General Assembly, June 11, 2004. http://www.un.org/french/ga/search/view_doc.asp?symbol=A/58/817&referer=http://www.un.org/french/reform/panel.html&Lang=E

Castells, Manuel. *The Rise of the Network Society*. Oxford: Blackwell, 2001.

———. "The Rise of the Fourth World." In David Held and Anthony G. McGrew (eds.), *The Global Transformations Reader*. Cambridge: Polity Press, 2003: 430-439.

Clarke, John N. and Geoffrey R. Edwards (eds.). *Global Governance in the Twenty-First Century*. Houndmills, Basingstoke, Hampshire; New York: Palgrave Macmillan, 2004.

Communications. Special section: "Media Governance: New Ways to Regulate the Media." *Communications–The European Journal of Communication Research* 32 (3). 2007: 323-362.

Communication Rights in the Information Society (CRIS). *Assessing Communication Rights: A Handbook*. 2005. http://www.crisinfo.org/pdf/ggpen.pdf

Dawkins, Kristin. *Global Governance: The Battle over Planetary Power.* New York: Seven Stories; London: Turnaround, 2003.

Dijk, Jan A.G.M. van. *The Deepening Divide: Inequality in the Information Society.* London: Sage, 2005.

Goldsmith, Jack L. and Tim Wu. *Who Controls the Internet? Illusions of a Borderless World.* New York: Oxford University Press, 2006.

Hackett Robert A., and Yuezhi Zhao (eds.). *Democratizing Global Media: One World, Many Struggles.* Lanham, MD: Rowman & Littlefield Publishers, 2005.

Hamelink, Cees. *The Politics of World Communication: A Human Rights Perspective.* London: Sage, 1994.

Held, David and Anthony McGrew (eds.). The *Global Transformations Reader: An Introduction to the Globalization Debate.* Malden, Mass.: Polity Press, 2000.

———. *Governing Globalization: Power, Authority and Global Governance.* Cambridge, UK: Polity Press; Malden, MA: Blackwell Publishers, 2002.

Hemmati, Minu. *Multi-Stakeholder Processes for Governance and Sustainability: Beyond Deadlock and Conflict.* London; Sterling, VA: Earthscan Publications, 2002.

Hewson, Martin and Timothy J. Sinclair (eds.). *Approaches to Global Governance Theory.* Albany, NY: State University of New York Press, 1999.

Hoffmann, Jeanette. "ICANN," in *Global Information Society Watch 2007.* Montevideo: Association for Progressive Communications and Third World Institute, 2007.

Hughes, Steve (ed.). *Global Governance: Critical Perspectives.* London: Routledge, 2002.

Jørgensen, Rikke Frank (ed.). *Human Rights in the Global Information Society.* Cambridge, Mass: MIT Press, 2006.

Kahler, Miles. *Networked Politics: Agency, Power, and Governance.* Ithaca, NY: Cornell University Press, 2009.

Kingdon, John. *Agendas, Alternatives and Public Policies.* New York: Harper-Collins Press, 1984.

Lessig, Lawrence. *Code and Other Laws of Cyberspace.* New York: Basic Books, 1999.

———. *Code Version 2.0* (2nd ed.). New York: Basic Books, 2006.

MacBride, Sean, et al. *Many Voices, One World: Towards a New, More Just, and More Efficient World Information and Communication Order.* Lanham, MD: Rowman & Littlefield, 2004.

MacLean, Don. "The Quest for Inclusive Governance of Global ICTs: Lessons from the ITU in the Limits of National Sovereignty." *Information Technologies and International Development* 1 (1), 2003: 1-18.

———. (Ed.). *Internet Governance: A Grand Collaboration.* New York: UN ICT Task Force, 2004.

Mattelart, Armand. *The Information Society: An Introduction.* London: Sage, 2003.

McDowell, Stephen D., Philip E. Steinberg and Tami K. Tomasello. *Managing the Infosphere: Governance, Technology and Cultural Practice in Motion.* Philadelphia: Temple University Press, 2008.

McKeon, Nora. The *United Nations and Civil Society: Legitimating Global Governance–Whose Voice?* London; New York: Zed, 2009.

McLaughlin, Lisa and Victor Pickard. "What is Bottom Up about Global Internet Governance?" *Global Media and Communication* 1 (3), 2005: 357-373.

Mueller, Milton L. *Ruling the Root: Internet Governance and the Taming of Cyberspace*. Cambridge, Mass: MIT Press, 2002.

Mueller, Milton L., Brenden N. Kuerbris, and Christiane Pagé. "Democratizing Global Communication? Global Civil Society and the Campaign for Communication Rights in the Information Society." *International Journal of Communication* 1, 2007: 267-296.

Nye, Joseph S., and John D Donahue. *Governance in a Globalizing World*. Cambridge, Mass.: Brookings Institution Press, 2000.

Ó Siochrú, Sean. "Finding a Frame: Towards a Transnational Advocacy Campaign to Democratize Communication." In Robert A. Hackett and Yuezhi Zhao (eds.), *Democratizing Global Media: One World, Many Struggles*. New York, Oxford: Rowman & Littlefield, 2005: 289-311.

———. "Global Media Governance as a Potential Site of Civil Society Intervention." In Robert A. Hackett and Yuezhi Zhao (eds.), *Democratizing Global Media: One World, Many Struggles*. New York, Oxford: Rowman & Littlefield, 2005: 205-221.

———. "Implementing Communication Rights." In Marc Raboy and Jeremy Shtern, *Media Divides: Communication Rights and the Right to Communicate in Canada*. Vancouver: UBC Press, 2010.

Ó Siochrú, Sean and Bruce Girard. *Global Media Governance: A Beginner's Guide*. Boulder, CO. & London: Rowman & Littlefield, 2002.

———. (Eds.). *Communicating in the Information Society*. Geneva: UNRISD, 2003.

Paré, Daniel J. *Internet Governance in Transition: Who is the Master of this Domain?* Lanham, MD: Rowman & Littlefield, 2003.

Pavan, Elena. *Transnational Communication Networks on Internet Governance. Mapping an Emerging Field*. Unpublished PhD Thesis, University of Trento, Italy, 2009.

Peyer, Chantal (ed.). *Who Pays for the Information Society? Challenges and Issues on Financing the Information Society*. PPP-Reperes (1) 05. Lausanne: Bread for All, 2005: 9-24.

Pierre, Jon and B. Guy Peters. *Governance, Politics, and the State*. New York: St. Martin's Press, 2000.

Raboy, Marc. "Communication and Globalization: A Challenge for Public Policy." In David Cameron and Janice Gross Stein (eds.), *Street Protests and Fantasy Parks: Globalization, Culture and the State*. Vancouver: UBC Press, 2003: 109-140.

Raboy, Marc and Aysha Mawani. "Are States still Important? Reflections on the Nexus Between National and Global Media and Communication Policy." In Andrew Calabrese and Claudia Padovani (eds.), *Communication Rights and Global Justice: Reflections on the Short History of a Social Movement*. Creskill, NJ: Hampton Press (forthcoming).

Raboy, Marc and Claudia Padovani. "Mapping Global Media Policy: Concepts, Frameworks, Methods." *Journal of Communication, Culture & Critique*, 3 (2), 2010: 150-169.

Raboy, Marc and Jeremy Shtern. *Media Divides: Communication Rights and the Right to Communicate in Canada*. Vancouver: UBC Press, 2010.

Reinicke, Wolfgang H., Francis Deng et al. *Critical Choices: The United Nations, Networks, and the Future of Global Governance*. Ottawa: International Development Research Centre, 2000.

Rittberger, Volker (ed.). *Global Governance and the United Nations System*. Tokyo; New York: United Nations University Press, 2001.

Rhodes, R.A.W. *Understanding Governance: Policy Networks, Governance, Reflexivity, and Accountability*. Buckingham; Philadelphia, PA.: Open University Press, 1997.

Sinclair, Timothy J. (Ed.). *Global Governance: Critical Concepts in Political Science*. London; New York: Routledge, 2004.

Schuppert, Gunnar Folke (ed.). *Global Governance and the Role of Non-State Actors*. Baden-Baden: Nomos. 2006.

Smith, Jackie, Charles Chatfield, and Ron Pagnucco (eds.). *Transnational Social Movements and Global Politics: Solidarity Beyond the State*. Syracuse, N.Y.: Syracuse University Press, 1997.

Väyrynen, Raimo (ed.). *Globalization and Global Governance*. Lanham, Md.: Rowman & Littlefield, 1999.

Webster, Frank. *Theories of the Information Society* (second edition). London: Routledge, 2002.

Wellman, Barry. "The Three Ages of Internet Studies: Ten, Five and Zero Years Ago." *New Media and Society* 6(1), 2004: 123–129.

Whitman, Jim. *The Limits of Global Governance*. London; New York: Routledge, 2005.

Wilkinson, Rorden (ed.). *The Global Governance Reader*. London; New York: Routledge, 2005.

Wilkinson, Rorden and Steve Hughes (eds.). *Global Governance: Critical Perspectives*. London; New York: Routledge, 2002.

Zittrain, Jonathan. *The Future of the Internet and How to Stop It*. New Haven, CT: Yale University Press, 2008.

Websites

Official Websites of International Institutions

2005 World Summit
http://www.un.org/Summit2005.

Commission on Science and Technology for Development (CSTD)
www.unctad.org/cstd

Digital Solidarity Fund (DSF)
http://www.dsf-fsn.org

Earth Summit
http://www.earthSummit2002.org

Global Alliance for ICT and Development (GAID)
http://www.un-gaid.org/

Global Knowledge Partnership
http://www.globalknowledge.org/WSIS2005/

ICANN
http://www.icann.org/

International Telecommunication Union (ITU)
http://www.itu.int/home/index-fr.html

> **ITU's Digital Opportunity Index (DOI)**
> http://ictlogy.net/bibciter/reports/projects.php?idp=627

IT for Change
www.ITforChange.net

Internet Governance Forum (IGF)
www.intgovforum.org

Millennium Development Goals (MDGs)
http://www.un.org/millenniumgoals/

Science & Technology for Development
http://stdev.unctad.org/index.html

UNESCO
www.unesco.org

>**UNESCO Observatory on the Information Society**
>http://portal.unesco.org/ci/ev.php?URL_ID=7277&URL_DO=DO_TOPIC&URL_SECTION=201&reload=1048272936

>**UNESCO and WSIS**
>www.unesco.org/wsis

>**UNESCO Institute for Statistics**
>http://www.uis.unesco.org

UNGIS: United Nations Group on the Information Society
http://www.ungis.org

United Nations Development Program (UNDP)
http://www.undp.org

Working Group on Internet Governance
www.wgig.org

>**Inventory of Public Policy Issues and Priorities**
>http://www.wgig.org/docs/inventory-issues.html

>**Working Papers**
>http://wgig.org/working-papers.html

>**Internet Governance Arrangements: Questionnaire**
>http://www.wgig.org/docs/Questionnaire.09.05.05.pdf

World Summit of Cities and Local Authorities on the Information Society
http://www.it4all-bilbao.org

WSIS Official Website
http://www.itu.int/wsis/

Civil Society & News Sites

Citizens' Summit on the Information Society
http://citizens-Summit.org/objectives.html

Civil Society Meeting Point
http://www.wsis-cs.org/

Civil Society Plenary Archives Page
http://mailman.greennet.org.uk/mailman/listinfo/plenary

Civil Society Bureau (CSB)
http://www.csbureau.info/contactinformation.htm

Conference of NGOs (CONGO)
http://www.ngocongo.org/index.php?what=news&g=12

Choike (Portal on Southern Civil Society Activities)
http://wsispapers.choike.org/

CRIS (Communications Rights in the Information Society)
www.crisinfo.org

Heinrich Böll Foundation
www.worldSummit2005.org

International Freedom of Expression eXchange (IFEX)
http://www.canada.ifex.org/en/

The Public Voice Project
www.thepublicvoice.org

ZDNet
http://www.zdnet.com/
http://www.zdnet.fr/actualites/Internet/0,39020774,39287378,00.htm

Business

Coordinating Committee of Business Interlocutors (CCBI)
http://www.iccwbo.org/businessatwsis/

Index

A
academic research skills, 249-50
accessibility issues, 19, 29, 76, 158, 240, 252
acronyms, list of, xiii-xvi
action lines, WSIS, 158, 162, 202-3, 217
activism, 222-23, 229, 231
Affirmation of Commitments, 197-200
AMARC (World Association of Community Radio Broadcasters), 221
Annan, Kofi, 163, 180
Antalya Plenipotentiary Conference, 121, 213
Arrangements for Accreditation, 31
Arrangements for Participation, 31
Asia-Pacific conference, 65-66
assistive technologies issues, 251-52
Association for Progressive Communications (APC)
  follow-up of ITCD by, 202
  grassroots group, 221, 223
  on ICANN MoU, 197
  on local IGFs, 188
  Phase II organizational hub, 170, 171
  TFFM concerns, 108

B
Banks, Karen, 52
BASIS (Business Activities to Support the Information Society), 233
Beaird, Dick, 144, 192
Belhassen, Souhayr, 51, 52-53
Bendrath, Ralf, 49-50, 52, 64, 75, 84, 89
Boltanski, Christophe, 75
Brief History of WGIG (MacLean), 132
Brown, Mark Malloch, 56
Buckley, Steve, 52, 111-12
Burch, Sally, 60
business websites, 268

C
Cammaerts, Bart, 169-70
Cardoso, Fernando Henrique, 225
Cardoso report, 229-31

Carpentier, Nico, 169-70
Carr, John, 186
Carson, Rachel, 223
Carvin, Andy, 51-52
caucuses
  creation/role of, 38-39
  evaluation of, 81-82
  WMWG recommendations for, 85, 86
  See also specific caucuses
Cerf, Vinton, 147
chairs, of CS Plenary, 85
Chair's Paper draft, on IG, 142
Champions Network, in GAID, 211
changes, Phase I, 14-16
charter, Internet Governance Caucus, 176-79
  See also Internet Governance Caucus (IGC)
Chatham House Rule, 181
Chief Executives Board for Coordination (CEB), 162-63
children
  information society and, 253-54
  online safety of, 186-87, 189, 196
Citizens' Summit on the Information Society, 74-76, 241
Civil Society Bureau (CSB)
  creation/mandate of, 35-36
  evaluation of, 81
  organizing Phase II and, 45
  reform of, 82-83
  re-structuring recommendations, 85, 87
  subgroup formation, 53, 80
Civil Society, Communication and Global Governance: Issues from the World Summit on the Information Society (Raboy and Landry), 1-2, 6
civil society (CS)
  continued evolution of, 255
  criticism of TFFM, 108
  definitions/guidelines for structures, 85-87
  definition/term usage, 1, 225
  endogenous structures, 36-39
  evaluation outcomes, 87-88

global communication governance and, 197–200, 221–24
global governance/multi-stakeholder participation, 27–28
goals for Tunis Phase, 237–38
human rights issues and, 26–27
ideal role in Internet governance, 193–94
implementation process concerns, 152–53
implementing lessons learned, 165–66
importance of media and, 30
inclusivity in information society, 29
increased participation in Phase II, 15
information society and, 17–18
intellectual property rights issues, 29–30
Internet governance issues and, 27
key documents produced, 259
legitimacy issue, 224–26
negotiating follow-up measures, 153–59
official structures, 34–36
participation in WSIS, 6
at Phase II. *See* Phase II, civil society at
positions on financing, 115–18
at PrepCom I, 49–53
at PrepCom III, Geneva, 70–73
presenting financing principles, 111–12
remobilizing for Phase II, 43–45
reorganization/evaluation, 59
representation/performance, 226–27
review of existing structures, 81–84
roles/responsibilities of, 245–46
rules/procedures for participation, 31–33
social justice/financing issues, 24–26
strategy/effectiveness, 227–29
Tunis Commitment of, 255–57
views on official negotiations, 64
websites, 267–68
WSIS participation issues, 244–46
Civil Society Division (CSD)
creation/mandate of, 34–35
dismantling of, 42
Civil Society Information Society Advisory Council (CSISAC), 209
civil society, Internet governance and IGF
critique of CS participation, 194–96
CS accomplishments, 197–200
dynamic coalitions of IGF, 184–85
establishment/anatomy of IGF, 180–82
focal CS interest in IG, 169–71
ideal role for CS participation, 193–94
IGC charter drafting/adoption process, 176–79
IGC growth/meeting CTG criteria, 172–74
IGC legitimacy/transparency issues, 174–76
IGF and multi-stakeholder global governance, 187–90
IGF high points/tensions, 185–87
IGF overview, 179
IGF structure/experience, 182–84
ITU threat to multi-stakeholderism, 189–90
revisionist multi-stakeholder model, 191–93
civil society participation, post-WSIS
in GAID, 211–12
ICANN and, 206–8
in ITU, 212–15
OECD and, 208–9
overview, 201–2
at UNESCO Convention, 204–5
WIPO and, 209–10
in World Social Forum process, 205–6
in WSIS implementation/follow-up, 202–4
Civil Society Plenary (CSP)
adopting reform plans, 70–71
approving CS structure mandates, 88
creation/role of, 36–37
evaluation of, 81
re-structuring recommendations, 85, 86
Clarke, Trevor, 210
Clerc, Alain, 106–7
closed club model, of global governance, 232–33, 234
collective memory, of civil society, 80
Commission on Science and Technology for Development (CSTD)
multi-stakeholder participation in, 245
open club model of global governance and, 234
revision of, 256
role of, 161–62
WSIS follow-up activities and, 203
communication governance
international context for, 3–4
Internet governance and. *See* Internet governance, at WSIS
principles and values of, 4–5
related topics on, 263–66
after WSIS. *See* global communication governance, post-WSIS

Communication Rights in the Information
  Society (CRIS) campaign
    engagement in World Social Forum, 205-6
    Internet governance and, 27, 169, 170, 171
    after Phase I, 201-2
    UNESCO Convention and, 204-5
Conference of Non-Governmental
  Organizations (CONGO)
    Civil Society Bureau and, 35
    distributing code of conduct, 88
Content and Themes Group (CTG)
    conditions for membership, 173-74
    creation/role of, 37-38
    evaluation of, 81
    re-structuring recommendations, 85, 86
continuity, of WSIS Phases, 16-17
conventional wisdom, on Internet
  governance, 198-200
Convention on the Elimination of All Forms
  of Discrimination against Women
  (CEDAW), 247
Convention on the Protection and Promotion of
  the Diversity of Cultural Expressions, 204-5
Coordinating Committee of Business
  Interlocutors (CCBI), 123, 126-27, 233
Council of Europe, 234
Country Code Top Level Domain (CCTLD)
    Internet principles for, 139
    as public good, 243
cross-cutting issues
    of both Phases, 28-30
    for CS participation, 44-45
    cultural/linguistic diversity as, 247-49
    ICT integration/poverty reduction, 102-3
CS Internet Governance Caucus
    charter drafting/adoption, 176-79
    controversies, 39
    creation of WGIG and, 132-33
    defining CS WSIS participation, 171
    growth/meeting CTG criteria, 172-74
    on ICANN MoU, 197
    legitimacy/transparency/democracy issues,
      174-76
    organizing information sessions, 142
    setting WGIG guidelines, 91
    Subcommittee A input and, 71
Cultural Diversity Convention, 201
cultural diversity issue, 4, 158, 204, 248-49
Currie, Willie, 202, 231-32

D

Declaration of Principles
  on Digital Solidarity Fund, 105-6
  on financing, 112-13
  follow-up/implementation of, 47
  on gender equality, 247
  at Geneva, 11
  human rights issues and, 241
  on Internet governance, 123-25, 127, 128,
    130-31
  on public financing, 22-23
  significance of, 218-19, 220
declarations, of WSIS, 259
de La Chapelle, Bertrand, 156, 192
democracy, multi-stakeholder, 229-32
Desai, Nitin, 58, 132, 182
development issues, of information society, 29
digital divide issue
  as central topic of WSIS, 16
  debates on, 61-62
  structural inequality and, 18-19
  WSIS response to, 104-5
Digital Opportunity Index, 165
Digital Solidarity Agenda, 22
Digital Solidarity Fund (DSF)
  CS support of, 116
  discussion of, 61-62
  financial support for, 240-41
  goals/establishment of, 25, 105-7
  launching of, 65
diplomatic immunity, for Tunis Summit, 59-60
disability-specific approach to ITC, 251-52
discrimination, gender-based, 246-47
documents, produced by CS, 259
Domain Name System (DNS)
  ICANN and, 120
  intersessional meeting proposals, 127
  management issue, 120-25
  private technical management of, 199
  US government principles and, 139-40
Doria, Avri, 96
Drake, Bill, 135, 137-39, 193, 214
dynamic coalitions (DCs), 184-85, 189
Dyson, Esther, 120

E

Economic and Social Council (ECOSOC)
  CSTD. *See* Commission on Science and

Technology for Development (CSTD)
  implementation/follow-up process and, 161
  open club global governance and, 234
education issues
  digital divide and, 19
  in information society, 29
  resources for ICT, 249-50
electronic pollution, 114
endogenous structures, of civil society
  caucuses/working groups, 38-39
    Civil Society Content and Themes Group, 37-38
    Civil Society Plenary, 36-37
    overview, 36
equity issues, 29
Esterhuysen, Anriette, 203
ethical dimensions, of WSIS, 254
European Civil Society Caucus, 111-12
European Union (EU), 22, 33, 61, 143-46
European WSIS Ministerial Meeting, 120-23
exclusion, from information society, 4, 22, 25-26, 29, 228
Extract from Draft Plan of Action, 128

F
figures, list of, ix
financial assistance, for WSIS participation, 44-45
financing, information society
  CS support of, 24-26
  discussed at PrepCom II, 61-62
  international cooperation, 22-23
financing issues, Phase II
  controversies, 102-4
  CS perspective on, 115-18
  CS principles on, 111-12
  CS summary statement, 239-41
  establishing DSF, 105-7
  overview, 101
  refusal of digital inclusion, 112-15
  response to digital divide, 104-5
  TFFM activities/final report, 107-11
follow-up mechanisms. *See* implementation/follow-up process
free software access, 248
Furrer, Marc, 129

G
G8 (Group of Eight), 234
G20 (Group of Twenty), 234
Gagné, Pierre, 42
Geiger, Charles, 42, 73, 213, 214
gender equality issues, 29, 246-47
Generic Names Supporting Organization (GNSO), 206-8
Geneva Plan of Action
  action lines, 158, 162, 202-3, 217
  adoption of, 11
  on Digital Solidarity Fund, 106
  implementation/follow-up of, 47, 150-52, 255
  information society financing, 23
  on Internet governance, 123-24, 128, 131
  Internet governance mandate, 21
  objectives, 103-4
  Phase II mandates, 13
Geneva Principle, 66, 238, 240, 242
Geneva Summit. *See* Phase I (Geneva Summit)
Global Alliance for ICT and Development (GAID), 211-12
global communication governance, post-WSIS
  CS, IG and IGF. *See* civil society, Internet governance and IGF
  CS involvement in, 197-200, 221-24
  CS participation, after WSIS. *See* civil society participation, post-WSIS
  multi-stakeholder. *See* multi-stakeholder global governance
  overview, 167
  See also global governance issues
global development, 3-4
Global Forum on Disability in the Information Society, 251
global governance issues
  civil society support of, 27-28
  CS accomplishments in, 197-200, 221-24
  CS summary report on, 244
  global governance models, 33-34, 232-35
  new precedent in, 42
  related topics on, 263-66
  WSIS contributions to, 217-21
  See also global communication governance, post-WSIS
Global Information Society Watch
  APC-sponsored, 202

INDEX

overviewing WIPO, 209-10
global public goods
   Brazil's views on, 124-25
   CCTLDs as, 243
   ICTs as, 116-17
   Internet access as, 240
   knowledge as, 110-12, 247-48
global social solidarity, 117-18
Golden Book, 159
Governmental Advisory Committee (GAC)
   open club global governance model, 234
   possible reform of, 147
Governmental Non-Governmental
   Organizations (GNGOs), 79
Gross, Robin, 209-10
Group of Eight (G8), 234
Group of Friends of the Chair (GFC)
   completing work, 68
   composition/working methods of, 53-55
   multi-stakeholder principle in, 89-90
   preparing August 2005 draft, 155-56
   proposal for stocktaking, 153-54
   supporting PrepComs work, 17
Group of Twenty (G20), 234
guidelines
   for CS structures, 85-87
   for WGIG, 90
Gurstein, Michael, 231

H
health information issues, 252
Human Rights in China (HRC), 73
human rights issues
   CS summary statement on, 241-42
   CS support of, 26-27
   as IGF agenda item, 185-86
   Internet governance and, 62
   PrepCom I caucus session on, 51-53
   violation at PrepCom III, Tunis, 74-76

I
ICT Design, 251-52
ICT Opportunity Index, 165
illiteracy statistics, 19, 101
implementation/follow-up process
   adoption of Tunis Agenda, 157-58
   allowing for NGO input, 160
   CS involvement in, 255-56

   CSTD role in, 161-62
   CS views/concerns, 152-53
   ECOSOC responsibility for, 161
   facilitators group for WSIS action lines, 162
   formation of CS working group, 156-57
   GFC proposal, 153-54
   GFC work and, 68
   interest/support for, 159
   ITU stocktaking procedure, 164
   lessons learned, 165-66
   national/regional/international level strategies, 163-64
   after Phase I, 23-24
   at PrepCom II, 63-64
   at PrepCom III, 70
   statistical indicators of, 164-65
   UNGIS role in, 162-63
   weakened GFC text, 155-56
   WSIS as test for, 149-52
impossibility, of Internet governance, 198-99
Industry Canada, 197
Information and Communication for
   Development (ICD), 111
Information and Communication
   Technologies (ICTs)
   children/young people and, 253-54
   as development tools, 240, 250
   education/research for, 249-50
   poverty reduction goals and, 102-3
   UN task force on, 211
   WSIS discussion of, 17-20
Information and Communication Technology
   for Development (ICTD)
   ACP focus on, 202
   CS perspective on, 116
   financing, 239-40
Information and Communication World
   Forum (ICWF), 205-6
information flow, 3
information society
   financing, 22-23 *See also* financing issues, Phase II
   proprietary technology and, 29-30
   struggle for inclusivity in, 29
   WSIS discussion of, 17-19
intellectual property rights issues, 29-30
intergovernmental Internet governance, 123-26
Intergovernmental Plenary, 49

interlude, first
    CS reorganization/evaluation, 59
    GFC composition/working methods, 54–55
    launching of WGIG, 57–58
    main issues during, 53
    regional conferences, 58
    stocktaking principles, 56
    TFFM during, 56–57
interlude, second
    completion of GFC work, 68
    launching of DSF, 65
    overview, 64–65
    regional conferences, 65–66
    WGIG final report, 66–67
International Bill of Human Rights, 227
International Chamber of Commerce (ICC), 233
international club model, of global governance, 232–33, 234
International Convention on the Rights of Persons with Disabilities, 252
international cooperation, on financing, 22–23
international institutions' websites, 266–67
international level, of implementation strategies, 158, 163–64
International Network for Cultural Diversity (INCD), 204–5
international technological innovations, 3–4
International Telecommunication Union (ITU)
    CS participation in, 212–15
    data from, 19–20
    Internet governance and, 121
    mandate to organize WSIS, 3
    PrepCom III resumed II and, 129
    stocktaking procedure of, 164
    threat to multi-stakeholderism, 189–90
Internet Architecture Board (IAB), 120
Internet Assigned Numbers Authority (IANA), 185–86
Internet Corporation for Assigned Names and Numbers (ICANN)
    creation of WGIG and, 135
    four models of IG and, 66–67
    functions of, 120
    GNSO improvement, 206–8
    international/intergovernmental control, 123–26
    Internet governance and, 27

MAG membership from, 180, 181
management issue, 120–23
oversight controversy, 139–40
private technical management and, 199
public overtures of, 147
root zone file control and, 70
US government oversight of, 197
Internet Engineering Task Force (IETF)
    NomCom Random Selection process, 176–77
    standards-making processes and, 120
Internet Governance Ad-Hoc Working Group, 127–28
Internet governance, at WSIS
    basic issues, 27
    creation of WGIG, 131–35
    CS summary statement on, 242–44
    in Declaration of Principles/Plan of Action, 130–31
    domain name management issue, 120–23
    EU new cooperation model, 143–46
    international vs. intergovernmental, 123–26
    intersessional meeting, 127–28
    language usage agreement, 129–30
    overview, 119
    Phase I and ICANN, 120
    Subcommittee A meeting, 140–42
    Tunis compromise, 147–48
    WGIG report, 135–40
Internet Governance Caucus (IGC)
    charter drafting/adoption process, 176–79
    controversies, 39
    creation of WGIG and, 132–33
    defining CS WSIS participation, 171
    growth/meeting CTG criteria, 172–74
    on ICANN MoU, 197
    legitimacy/transparency/democracy issues, 174–76
    organizing information sessions, 142
    setting WGIG guidelines, 91
    Subcommittee A input, 71
Internet Governance Forum (IGF)
    CS participation in, 242–43, 255
    dynamic coalitions, 184–85
    establishment/anatomy of first, 180–82
    high points/tensions, 185–87
    information sessions on, 142
    meeting themes: 2006-2009, 183
    multi-stakeholder global governance and,

187-90
  multi-stakeholder participation in, 67
  open club global governance model, 234
  overview, 179
  structure/experience, 182-84
  Tunis compromise and, 147-48
Internet governance (IG)
  civil society support of, 27
  conventional wisdom on, 198-99
  critique of CS involvement, 194-96
  CS accomplishments in, 197-200
  definition, 66
  four models of, 66-67
  ideal role for CS in, 193-94
  key public policy issues, 135
  launching of WGIG, 57-58
  in Plan of Action draft, 123-24
  PrepCom I focus, 47
  PrepCom II discussion of, 62-63
  root zone file control, 70
  technical vs. political approach to, 20-21
  WGIG definition of, 136
  WGIG models for reform, 137
  after WSIS. *See* civil society, Internet governance and IGF
  WSIS contributions to, 220
Internet Protocol (IP) address assignment, 123, 125
Internet Society (ISOC)
  creation of WGIG and, 135
  intervention of, 125-26
  MAG membership from, 180, 181
Internet websites
  blocking of, 75
  business, 268
  of CS and news, 267-68
  of international institutions, 266-67
intersessional meeting proposals, 127-28
issues, WSIS
  children/young people, 253-54
  culture/knowledge/public domain, 247-49
  education/research/practice, 249-50
  ethical dimensions, 254
  gender equality, 246-47
  health information, 252
  media, 250-51
  overview, 17-20
  Phase II. *See* Phase II, issues/outcomes

universal design/assistive technologies, 251-52
IT for Change
  connecting people with issues, 223
  on ICANN MoU, 197
ITU Plenipotentiary Conference, 189-90

J
Joint Projects Agreements (JPA), 197

K
Karklins, Janis, 32, 49, 68, 90, 95-96, 155, 214
key civil society documents, 259
Khan, Masood, 231
Kingdon, John, 235
Kleinwächter, Wolfgang, 66, 121, 197
knowledge
  as global public good, 110-12, 247-49
  societies, UNESCO on, 18
Knowledge Ecology International, 210
Kummer, Markus, 46, 128, 129-30, 131, 132, 182

L
Latin American and Caribbean conference, 65-66
Latin American Regional Ministerial Conference, 122
legacy, of WSIS, 217-21
legitimacy, for civil society, 224-26
libertarian model of global governance, 232, 233
linguistic diversity, 4, 29, 62, 158, 248
listserv, Internet Governance Caucus, 178-79, 181
London School of Economics (LSE) Public Policy Group, 206-8
long march through the institutions, global governance model, 233, 234-35
Love, James, 210
Lucero, Everton, 186

M
MacBride Commission debates, 221-22
MacLean, Don, 132, 135
MAG (Multi-stakeholder Advisory Group)
  composition/role in IGF, 180-82
  CS representatives, 176-77

media issues
  CS summary statement on, 250-51
  importance of traditional, 30
mediation, for CS disputes, 85
Memoranda of Understanding (MoU), 197, 199
Ménard, Robert, 75
meta-issues, 222
Michael, Alun, 192
Millennium Development Goals (MDGs), 252
  contributing to, 16
  targeted issues of, 4, 102-3, 217
Monterrey Consensus, 240
Much More Could Have Been Achieved
  children/young people issues, 253-54
  CS Tunis Commitment, 255-57
  on culture/knowledge/public domain, 247-49
  drafting of, 77
  on education/research/practice, 249-50
  on ethical dimensions of WSIS, 254
  on gender equality, 246-47
  on global governance, 244
  on health information, 252
  on human rights, 241-42
  on Internet governance, 242-44
  introduction, 237-39
  on media, 250-51
  on participation, 244-46
  significance of, 219
  on social justice/financing, 239-41
  on universal design/assistive technologies, 251-52
Mueller, Milton, 134, 169, 170, 171, 181-82, 192-93
multilingualism, 189
Multi-stakeholder Advisory Group (MAG), 176, 180, 189
multi-stakeholder global governance
  CS exclusion from drafting groups, 94-97
  CS legitimacy issue, 224-26
  CS representation/performance, 226-27
  CS strategy/effectiveness, 227-29
  democracy in, 229-32
  GFC demonstration of, 89-90
  global communication governance and CS, 221-24
  IGF and, 187-90
  implementing, 33-34
  models of, 232-35

overview, 217
Phase II campaign for, 88-89
revisionist model, 191-93
WGIG/TFFM demonstrations of, 90-94
WSIS contributions to, 217-21
multi-stakeholder participation
  civil society support of, 27-28
  continuance in Phase II, 16
  defending at PrepCom I, 49-50
  in Internet governance forum, 67
  in ITU, 212-15
  ITU threat to, 189-90
  reaffirmation of, 44
  shared responsibility of, 14-15
  for stocktaking measures, 153-54
  in WSIS, 3, 5

N
national level, of implementation strategies, 157-58, 163-64
National Telecommunications and Information Administration (NTIA), 197
Networked Information and Communication Technologies (NICTs), 110-11
new cooperation model, for Internet governance, 143-46
news websites, 267-68
Nominating Committee, 133
Nominations Committee (NomCom)
  Random Selection process, 176-77
Non Commercial Users Constituency (NCUC)
  creation of WGIG and, 133-34
  GNSO improvement and, 206-8
Non-Governmental Organizations (NGOs), 212-13

O
observer status, in ITU, 212-13
Official Development Assistance (ODA), 240
official structures, civil society
  Civil Society Bureau, 35-36
  Civil Society Division, 34-35
online child protection, 186-87, 189, 196
Operational Part of the Final Document/Tunis Agenda for Action/Tunis Plan of Implementation as of January 2005, 60-61
organizational structure, in Phase I, 16-17

Organization for Economic Co-operation and Development (OECD)
  closed club model of global governance and, 234
  CS participation in, 208-9
  IGF and, 188
  Internet governance and, 20
organization, of WSIS, 63-64
outcomes, Phase II. *See* Phase II, issues/outcomes
outreach, to CS constituencies, 256-57

P
Partnership on Measuring ICTs for Development, 165
Peake, Adam, 96
people-centred development
  civil society support of, 24-26
  CS summary statement on, 239-41
Phase I (Geneva Summit)
  goal of, 14
  implementation/follow-up, 150
  issues raised by CS, 28-30
  process overview, 12
  unresolved issues, 11, 20-24
Phase II, civil society at
  definitions/guidelines for CS structures, 85-87
  evaluation outcomes, 87-88
  multi-stakeholder global governance campaign. *See* multi-stakeholder global governance
  overview, 79-81
  review of existing structures, 81-84
Phase II, issues/outcomes
  financing, of information society. *See* financing issues, Phase II
  implementation/follow-up. *See* implementation/follow-up process
  Internet governance. *See* Internet governance, at WSIS
  overview, 99-100
Phase II, preparatory framework
  detailed timetable, 42
  first interlude, 53-59
  organization, 45-46
  outline of process, 48
  overview, 41-43

PrepCom I. *See* Preparatory Committee (PrepCom) I
PrepCom II. *See* Preparatory Committee (PrepCom) II
PrepCom III. *See* Preparatory Committee (PrepCom) III, Geneva; Preparatory Committee (PrepCom) III, Tunis
reconstruction/remobilization, 43-45
second interlude, 64-68
Tunis Summit, 76-77
Phase II (Tunis Summit)
  characteristics of, 14-16
  issues raised by civil society, 24-30
  process overview, 12
Plan of Action. *See* Geneva Plan of Action
Plenary. *See* Civil Society Plenary (CSP)
Plenipotentiary Conference, 121, 213
policy entrepreneur, 235
political agenda
  continuity of Phase II, 16
  CS legitimacy and, 226
  failure of WSIS, 219
  Internet governance and, 20-21
  for Phase I, 14
  of Tunis government, 15-16, 51-52
  WSIS objectives, 18
political chapeau statement
  basis for negotiation, 60-61, 112
  goals of, 55
  at PrepCom III, 70, 74
Political Chapeau/Tunis Commitment as of 11 January 2005, 60
poverty reduction/eradication, 4, 102-3, 114, 211, 240
Preparatory Committee (PrepCom) I
  civil society at, 49-53
  outline of preparatory process, 48
  overview, 47-49
Preparatory Committee (PrepCom) II
  CS views on official negotiations, 64
  financing/DSF discussions, 61-62
  GFC proposal for stocktaking, 153-54
  Internet governance, 62-63
  overview, 59-61
  WSIS implementation/follow-up/organization, 63-64
Preparatory Committee (PrepCom) III, Geneva

CS participation in, 70-73, 244-45
Internet governance/control, 70
overview, 68-69
political chapeau/follow-up, 70
Preparatory Committee (PrepCom) III, Tunis
human rights hypocrisy, 74-76
overview, 73-74
stakeholder participation in, 94-97
preparatory framework for Phase II. *See* Phase II, preparatory framework
private technical management, of DNS, 199
public domain issues, 29-30, 247-49
public interest principle, 198-200
Publicly Verifiable NomCom Random Selection process, 176-77
public private partnership (PPP), 112, 125, 143

R
regional conferences
first interlude, 58
second interlude, 65-66
regional level, of implementation strategies, 157-58, 163-64
regulations, for multi-stakeholder participation, 5-6
remobilization, of WSIS, 43-46
Reporters Without Borders, 75
representation, of international civil society, 226-27
research skills issue, 249-50
Resolution 141, 213-14
resolution A/RES/56/183, 2-3, 102
resolution A/RES/57/270 B, 6-7, 149-50
revisionist multi-stakeholder model, 191-93
Rice, Condoleezza, 146
root zone files, 70, 136, 143-44
Rules of Procedure of the Preparatory Committee, 31-33

S
Sammasékou, Adama, 32, 49, 214
Schmid, Samuel, 76
second generation of WSIS CS, 170
sector membership, in ITU, 212-13
self-regulation model of global governance, 232, 233-34
Shaping Information Societies for Human Needs, 5, 219, 238

Silent Spring (Carson), 223
Singh, Parminder Jeet, 192-93, 211
social justice issues, 24-26, 239-41
social media platforms, 187
social movements, 221-23, 235
social network analysis, of WSIS CS, 170
soft power, 187
software, free vs. proprietary, 248
Souter, David, 203
stakeholder participation
in ITU, 214
organizing Phase II, 45-46
at PrepCom III, 94-97
rules/procedures for, 31-33
See also multi-stakeholder participation
statistical indicators, 164-65
stocktaking
establishing principles, 56
functions of, 164
ITU report on, 202
of WSIS implementation, 151-52
of WSIS objectives, 23-24
See also implementation/follow-up process
Straw, Jack, 146
Subcommittee A
compromise, at Tunis, 147-48
on Internet governance, 140-42
new cooperation model, 143-46
Subcommittee B
function of, 69, 140
preparatory process of, 74
working group on, 156-57
subcommittee meetings
access to, 32
at PrepCom III, 71-72, 74
stakeholder participation in, 94-96, 212
"summit of solutions", 23, 43, 150, 239
Swiss Federal Office of Communications, 197
Switzerland, preparatory process at, 13

T
tables, list of, xi
Tarmizi, Mohamed Sharil, 147
Task Force on Financial Mechanisms (TFFM)
activities/final report, 107-11
consideration of report by, 47
December 22, 2004 report, 61-62
information society financing and, 23

• INDEX • 279

launching/mandate of, 56-57
multi-stakeholder principle in, 90-94, 108
output for Phase II, 13-14
technological innovations, 3
technological transfers, 114
themes, of IGF meetings, 183
themes, WSIS, 17-20
  See also issues, WSIS
top-down institutional model of global governance, 232-33, 234
Touré, Hamadoun, 190
tripartite model, of global governance, 191-93
Tunis Agenda for the Information Society
  adoption of, 13, 77
  on child protection, 253
  CS critique of, 243
  on Digital Solidarity Fund, 106
  on financing issues, 113-14, 240
  on implementation/follow-up, 157-58
  implementation/follow-up of, 255
  importance of, 219
  on universal design concepts, 251-52
Tunis Commitment
  adoption of, 13, 77
  civil society's, 255-57
  on financing issues, 240
  on gender equality, 247
  human rights issues and, 241
  importance of, 219
Tunisian authorities, and human rights violations, 74-76
Tunis Summit
  CS goals for, 237-38
  CS participation in. *See* Phase II, civil society at
  designated topics for, 13-14
  issues/outcomes. *See* Phase II, issues/outcomes
  overview, 76-77
  political climate effect on, 15-16
  preparatory process. *See* Phase II, preparatory framework
  See also Phase II (Tunis Summit)
Twomey, Paul, 128
two phases, of WSIS
  changes in Phase II, 14-16
  continuity with Phase I, 16-17
  CS participation rules/procedures, 31-33

CS Phase II themes, 24-28
CS structures, 34-39
CS themes in both Phases, 28-30
defining phases of, 13-14
implementing multi-stakeholder model, 33-34
overview, 9
Phase I unresolved issues, 20-24
preparatory process, 11-13
from structure to substance, 39
themes/issues overview, 17-20

U

undesirability, of Internet governance, 199
United Nations Chief Executives Board for coordination (UNCEB), 162-63
United Nations Department of Economic and Social Affairs (UNDESA), 154
United Nations Development Program (UNDP), 93
  managing TFFM, 108
United Nations Economic and Social Council. *See* Economic and Social Council (ECOSOC)
United Nations Educational, Scientific and Cultural Organization (UNESCO)
  Convention issues, 204-5
  on illiteracy, 19, 101
  on knowledge societies, 18
  MacBride Commission debates, 221-22
United Nations General Assembly (UNGA)
  Ad Hoc Working Group of, 6, 149
  resolution A/RES/56/183 and, 2-3
  resolution A/RES/57/270, 6-7
  satisfying demands of, 41
United Nations Group on the Information Society (UNGIS)
  formation of, 159
  mandate of, 162-63
United Nations ICT task force, 211
United Nations Millennium Development Goals. *See* Millennium Development Goals (MDGs)
United Nations previous summits, 217
United Nations Secretary General (UNSG), 148
United Nations Security Council Resolution 141, 213-14
United Nations (UN), 58, 256
United Nations Working Group on Internet

Governance. *See* Working Group on Internet Governance (WGIG)
Universal Declaration of Human Rights (UDHR), 250
Universal Design Concept, 251-52
unresolved issues, of Phase I
    information society financing/international cooperation, 22-23
    Internet governance, 20-21
    stocktaking/implementation, 23-24
US delegation, on new cooperation model, 144-46
US Department of Commerce (US DOC), 120, 197
US government, Internet principles, 139-40

V

Vienna World Conference on Human Rights, 240-41

W

websites
    blocking of, 75
    business, 268
    of CS and news, 267-68
    of international institutions, 266-67
"We Seize" event, at Geneva, 241
West Asia Ministerial Conference, 122
women's empowerment, 246-47
Working Group on Internet Governance (WGIG)
    civil society support of, 27
    consideration of report by, 47
    creation of, 131-35
    CS participation in, 244
    launching of, 57-58
    listing CS roles/responsibilities, 245-46
    mandate of, 21
    models for institutional reform, 137
    multi-stakeholder principle in, 90-94
    multi-stakeholder principle reflected in, 50
    output for Phase II, 14
    preparatory activities of, 46
    publishing final report, 66-67
    report on key issues, 135-40
    responses to report, 140-42
Working Group on Subcommittee B, 156-57
working groups
    creation/role of, 38-39
    evaluation of, 81-82
    formed during Phase II, 15
    WMWG recommendations for, 85, 86
Working Methods Working Group (WMWG)
    goals of, 59, 80
    launching of, 53
    organizational audit by, 83-84
    proposals for reform, 85
World Association of Community Radio Broadcasters (AMARC), 221
World Bank (WB), 108
World Conference on International Telecommunications (WCIT), 190, 214
World Intellectual Property Organization (WIPO)
    CS participation in, 209-10
    IGF and, 187-88
World Social Forum (WSF), 205-6
World Summit on the Information Society
    on Digital Solidarity Fund, 106
    on financing issues, 240-41
    results of, 218
World Summit on the Information Society Executive Secretariat (WSIS ES)
    appointing head of, 46
    Civil Society Division in, 34
    implementing Plan of Action and, 151-52
World Summit on the Information Society (WSIS)
    declarations of, 259
    general analyses of, 259-63
    implementation/follow-up, 202-4
    as implementation/follow-up test, 149-52
    legacy of, 217-21
    multi-stakeholder participation in, 5-6
    Phase I. *See* Phase I (Geneva Summit)
    phase I, 1-2, 4, 6-7
    Phase II. *See* Phase II (Tunis Summit)
    process, 12
    themes/issues overview, 17-20
    two phases of. *See* two phases, of WSIS
    unique elements of, 3, 7
World Telecommunication Policy Forum (WTPF), 213-14
World Trade Organization (WTO), 201
WSIS action lines. *See* action lines, WSIS
WSIS Declaration of Principles. *See*

Declaration of Principles
WSIS Phase I. *See* Phase I (Geneva Summit)
WSIS Phase II. *See* Phase II (Tunis Summit)

Y
young people, and ICTs, 253–54